D0417184

For the updated syllabus

Biology

for Cambridge IGCSE®

Third edition

Revision Guide

Ron Pickering

Oxford and Cambridge
leading education together

OXFORD
UNIVERSITY PRESS

OXFORD
UNIVERSITY PRESS

Great Clarendon Street, Oxford, OX2 6DP, United Kingdom

Oxford University Press is a department of the University of Oxford. It furthers the University's objective of excellence in research, scholarship, and education by publishing worldwide. Oxford is a registered trade mark of Oxford University Press in the UK and in certain other countries

British Library Cataloguing in Publication Data
Data available

978-0-19-830872-0

10 9 8 7 6 5 4 3

Paper used in the production of this book is a natural, recyclable product made from wood grown in sustainable forests. The manufacturing process conforms to the environmental regulations of the country of origin.

Printed in Great Britain by Bell and Bain Ltd., Glasgow

Acknowledgements

® IGCSE is the registered trademark of Cambridge International Examinations.

The publisher would like to thank Cambridge International Examinations for their kind permission to reproduce past paper questions.

Cambridge International Examinations bears no responsibility for the example answers to questions taken from its past question papers which are contained in this publication. Unless otherwise indicated, the questions, example answers, marks awarded and comments that appear in this book and CD were written by the author. In examination, the way marks would be awarded to answers like these may be different.

The publishers would like to thank the following for permissions to use their photographs:

Cover image: Gary Yim/Shutterstock

p155: Ron Pickering; p156: Ron Pickering

Contents

If you have already read about the brain you might remember that an important function of the cerebral hemispheres is to act as an *integration centre* – input of information is compared with previous experience and an appropriate action is taken. As you prepare for an examination you will be inputting factual material and skills and you will be hoping that, when you're faced with examination papers, you will be able to make the correct responses! The effort you make in *revision* and your willingness to *listen to advice on techniques* will greatly affect your likelihood of success, as outlined here.

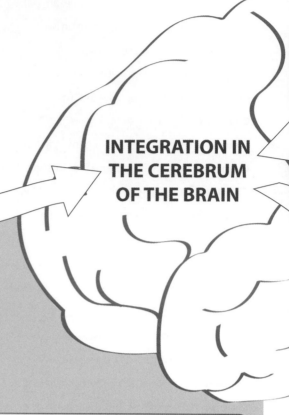

INTEGRATION IN THE CEREBRUM OF THE BRAIN

1 REVISION

There is no one method of revising that works for everyone. It is therefore important to discover the approach that suits you best. The following rules may serve as general guidelines.

BE PREPARED

Leaving everything until the last minute reduces your chances of success. Work will become more stressful, which will reduce your concentration. There are very few people who can revise everything 'the night before' and still do well in an examination the next day.

PLAN YOUR REVISION TIMETABLE

You need to plan your revision timetable some weeks before the examination and make sure that your time is shared suitably between all your subjects.
Once you have done this, follow it – don't be side-tracked. Stick your timetable somewhere prominent where you will keep seeing it – or better still, put several around your home!

RELAX

Concentrated revision is very hard work. It is as important to give yourself time to relax as it is to work. Build some leisure time into your revision timetable.

GIVE YOURSELF A BREAK

When you are working, work for about an hour and then take a short tea or coffee break for 15 to 20 minutes. Then go back to another productive revision period.

KEEP TRACK

Use checklists and the relevant examination board syllabus to keep track of your progress. Mark off topics you have revised and feel confident with. Concentrate your revision on things you are less happy with.

MAKE SHORT NOTES, USE COLOURS

Revision is often more effective when you do something active rather than simply reading material. As you read through your notes and textbooks make brief notes on key ideas. If this book is your own property you could highlight the parts of pages that are relevant to the specification you are following. Concentrate on understanding the ideas rather than just memorizing the facts.

PRACTISE ANSWERING QUESTIONS

As you finish each topic, try answering some questions. There are some in this book to help you (see Chapter 2). You should also use questions from past papers. At first you may need to refer to notes or textbooks. As you gain confidence you will be able to attempt questions unaided, just as you will in the exam.

2 KNOW WHAT TO DO

LEARN THE KEY WORDS

Name: the answer is usually a technical term (mitochondrion, for example) consisting of no more than a few words. **State** is very similar, although the answer may be a phrase or sentence. **Name** and **state** don't need anything added, i.e. there's no need for explanation.

Define: the answer is a formal meaning of a particular term, i.e. 'what is it?'
What is meant by...? is often used instead of **define**.

List: you need to write down a number of points (each may only be a single word) with no need for explanation.

Describe: your answer will simply say what is happening in a situation shown in the question, e.g. 'the temperature increased by 25°C' – there is no need for explanation.

Suggest: you will need to use your knowledge and understanding of biological topics to explain an effect that may be new to you. You might use a principle of enzyme action to suggest what's happening in an industrial process, for example. There may be more than one acceptable answer to this type of question.

Explain: the answer will be in extended prose, i.e. in the form of complete sentences. You will need to use your knowledge and understanding of biological topics to write more about a statement that has been made in the question or earlier in your answer. Questions like this often ask you to **state and explain**, and answers often include the word **because**.

Calculate: a numerical answer is to be obtained, usually from data given in the question. Remember to:
- give your answer to the correct number of significant figures, usually two or three
- give the correct unit
- show your working.

3

IN THE EXAMINATION

- Check that you have the correct question paper! There are many options in some syllabuses, so make sure that you have the paper you were expecting.
- Read through the whole paper before beginning. Select the questions you are most comfortable with – there is no rule that says you must answer the questions in the order they are printed!
- Read the question carefully – identify the key word (why not underline it?).
- Don't give up if you can't answer part of a question. The next part may be easier and may provide a clue to what you should do in the part you find difficult.
- Check the number of marks allocated to each section of the question.
- Read data from tables and graphs carefully. Take note of column headings, labels on axes, scales, and units used.
- Keep an eye on the clock – perhaps check your timing after you've finished 50% of the paper.
- Use any 'left over' time wisely. Don't just sit there and gaze around the room. Check that you haven't missed out any sections (or whole questions! Many students forget to look at the back page of an exam paper!). Repeat calculations to make sure that you haven't made an arithmetical error.

4 Success!

Using this study guide

Each section is arranged to help you to learn facts and to understand their importance, at whatever level you are studying!

The external assessment of IGCSE Biology (syllabus number 0610) can be made at two levels.

At **core** level you will study only the sections of the syllabus identified as **core**, and sit two written papers (numbers 1 and 3) as well as taking a practical assessment (Paper 5 or 6). The highest grade which you can be awarded for core papers is a grade C.

At **extended** level you will study both the **core** and the **supplement** sections in the syllabus. The two written papers are identified as Papers 2 and 4 (you will also take Practical Assessment 5 or 6), and can be awarded up to grade A*.

What are the differences between *core* and *supplement*: Paper 1 and 3 or Paper 2 and 4?

The core and the supplement papers are different in both the knowledge or facts you need, and in the skills you need to use in applying this knowledge.

- What extra *knowledge* is required for the supplement?

  The knowledge is quite distinct from the core e.g. *mechanism of* active transport, amino acid sequence and protein structure, DNA structure, limiting factors in photosynthesis, *explanation of* cholera, *details of* immune response, control of blood glucose, nitrogen cycle.
  You can find what is required for the supplement in the right hand column of your syllabus.

- What extra *skills* are required for the supplement?

  Generally questions on Papers 2 and 4 require *deeper understanding* of material that is covered in the core e.g. chemical digestion, negative feedback, kidney function, advantages and disadvantages of sexual and asexual reproduction. Papers 2 and 4 generally also need more *application* of knowledge.

The diagrams in each chapter of this study guide contain a lot of facts. Supplement material needed only for the extended papers is made obvious throughout with a wavy line so make sure to pay attention to the material that you need to cover.

By studying these revision diagrams you should have covered the *facts* that you need to answer questions on your IGCSE papers. You will do better on these papers if you practise using these facts so, following each of the revision pages, you will also find a series of examination-style questions.

Don't forget: candidates for the **extended** papers need to attempt *all* of the questions, i.e. those *with* the wavy line *and* those without it!

Crosswords and matching definitions exercises help with vocabulary and with definitions

Some candidates may not have English as a first language and may find some of the specific words in the syllabus challenging – perhaps difficult to define or difficult to spell. Crosswords can help with these problems as the correct answers can only be entered if:

- You can recognise the word from its definition, and
- You can spell the word correctly.

Many of the sections in this revision guide include a crossword as a summary of that part of the syllabus.

Other sections in the study guide have 'matching definitions' tests to act as summaries. These are very useful for checking that you can correctly identify biological terms from their definitions.

Answers

At the end of the book are the answers to all of these practice questions. As you check *your* answers against these, make sure that you have included all of the specific biological words (the best answers include suitable scientific vocabulary) and, if necessary, the units for any quantities used in calculations.

On the following pages some sample examination questions are answered. As you study them try to notice the following key points:

- what are you told to do and has the answer done it? Key words to look for are: **state; describe; explain; suggest; predict; calculate**
- the number of marks offered must be matched by the number of points made
- can you recognize what part of the syllabus is covered by the question? This will guide you to the likely answer
- some questions cover complicated material but actually contain most of the information you need to gain full marks!

1. The graph below shows the volume of air in the lungs of a person measured over a period of time.

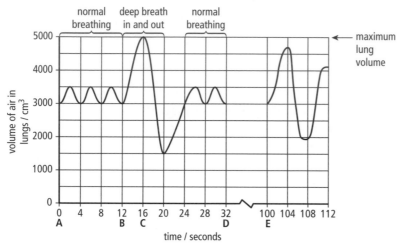

a. i.   With reference to the graph calculate, in breaths per minute, the rate of normal breathing between **A** and **B**. Show your working.

Don't forget! You will get a mark for correct working even if you make a mistake with the answer.

3 complete breaths in 12 seconds

So $3 \times \frac{60}{12}$ breaths in 60 seconds (ie 1 minute)

$= 3 \times 5 = 15$

..........15.......... breaths per minute (2)

Don't forget the units!

ii.   State the volume of air remaining in the lungs after the deep breath out.

1500 cm³

(1)

iii. Explain how the intercostal muscles are involved in breathing from time **B** to time **C**.

Explain is more than describe. Can your answer begin with the word 'because'?

Volume increases because external intercostal muscles

contract (1 mark) and lift the rib edge upwards

and outwards (1 mark)

(2)

At time **D**, the person performed one minute of vigorous exercise.

b. i.   On the graph starting at time **E**, continue the graph to show the person's breathing pattern after this exercise.   (2)

Notice this – many students don't!

ii.   Explain why the breathing pattern changes after a period of exercise.

Person should be recovering from deeper, faster breathing.

Because

- Less oxygen required/less CO_2 to be excreted
- Fewer, shallower breaths will be enough

(3)

(Total 10 marks)

Cambridge IGCSE Biology 0610 Paper 3 Q5 June 1998

2. a.   State **three** normal functions of a root.

1. To anchor the plant in the soil

2. To absorb water from the soil

3. To absorb mineral ions from the soil

(3)

The diagram below shows a mangrove tree growing in a swamp.

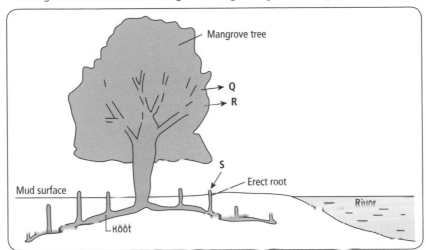

The roots of the mangrove are specially modified to overcome the fact that air spaces in the soil are always filled with water. In other respects, the roots are normal in structure and function.

Arrows **Q**, **R** and **S** represent the movement of gases into and out of the tree during the day. ——————————— So light is available for photosynthesis

b.   i.   Name gases **Q** and **R** and, for each gas, state the process in the tree which produces it.

Gas **Q** Oxygen .... process photosynthesis ....

Gas **R** Carbon dioxide .... process respiration ....

You need **both** answers to gain the mark—the gas must match the process

(2)

ii.   Name gas **S** and state in which process it is used in the tree. ——— Roots don't photosynthesize

Gas **S** Oxygen .... process respiration ....

(1)

c.   Name the unusual response being shown by the erect roots of the tree.

Positive phototropism (or negative geotropism)

(1)

Active transport is a process which occurs in most plant roots.

d.   Suggest why mangrove roots may have difficulty in carrying out this process. ——— Your answer should refer to some biological knowledge

• Active transport requires energy from respiration

• Roots can't obtain much oxygen for aerobic respiration

(2)

(Total 9 marks)

Cambridge Biology 0610 Paper 3 Q2 June 1998

3. The table below shows how three alleles control the ABO human blood group system.

ALLELE	EFFECT
I^A	Causes the production of antigen A on red blood cells
I^B	Causes the production of antigen B on red blood cells
I^O	Does not cause the production of antigens on red blood cells

When I^A and I^B are inherited together they are equal in their effect. When I^O is inherited — Important information which will be used later in the question
with either I^A or I^B it is recessive.

a. Complete the diagram below by filling in the blank circles to show the possible
inheritance of the ABO blood group system in Mr and Mrs Shah's children.

A clear **instruction** – no need for any **explanation** here

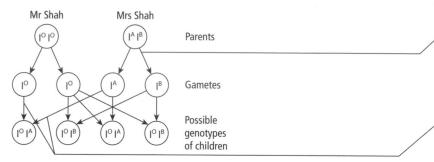

At meiosis (reduction division) only one of each pair of alleles from the diploid parent can be passed to any single haploid gamete

At fertilisation two haploid gametes fuse to form a diploid zygote

b. When I^A and I^B are inherited together they both show their effects in the
phenotype. Name the term used to describe this.
Co-dominance

Important! No call for any explanation

Define the term phenotype (1)

The total of the features of characteristics of an organism

Sex is determined by the X or Y chromosome carried in the male gamete. There are equal numbers of 'X' and 'Y' gametes.

c. i. What is the probability of the eldest child being female? (1)

50% or ½ or 1 in 2

(1)

ii. What is the probability of the eldest child being female with blood group B?

½ (being female) x ½ (being B) = ¼ or 25% or 1 in 4

(1)

N.B. **not** 1:4, as this means 1 in 5 or 20%

d. Sometimes a patient needs a blood transfusion. It is very important that the
blood given to the patient does not contain any antigen which is not in the
patient's own red blood cells. Which parent can donate blood safely to any of
the possible children?

Mr Shah (Since I^O blood has no damaging antigens)

(1)

(Total 5 marks)

4. a. Select the correct term from the list below and write it in the box next to its description.

— a very clear instruction

allele   dominant   gene   genotype   heterozygous

homozygous   phenotype   recessive

DESCRIPTION	TERM
a form of a gene that always has its effect when it is present	dominant
a form of a gene that codes for one of a pair of contrasting features	allele
an organism having two different forms of a gene for a particular feature	heterozygous
the alleles that an organism has in its chromosome	genotype

(4)

b. Two red flowered plants were crossed. The seeds produced were germinated and grew into 62 white flowered plants and 188 red flowered plants.

  i.   Which flower colour is controlled by the recessive form of the gene?

  white (the '1' in the 3:1)

(1)

  ii. Using the symbols **R** and **r**, construct a genetic diagram to explain the results of this cross.

Parents   Red   ×   Red
Genotype   Rr   Rr
Gametes   (R) (r)   (R) (r)

— You can't get this wrong if you use this 'square'

Offspring genotype

	(R)	(r)
(R)	RR	Rr
(r)	Rr	rr

Phenotype   Red   Red   Red   White
                    3   1

— don't forget to give the phenotypes

(4)

  iii. One of the white flowered offspring was crossed with a red flowered offspring. Predict the two possible ratios of red and white flowered plants that their seeds would produce.

— using your biological knowledge

1. All red (if red parent is homozygous – RR)

2. 1 red : 1 white (if red parent is heterozygous – Rr)

(2)

(Total 11 marks)

Cambridge Biology 0610 Paper 2 Q5 June 2005

5. The diagram below shows a food web from the Antarctic.

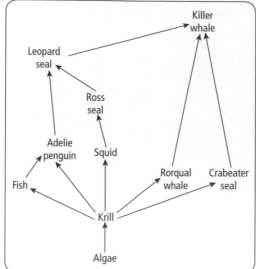

a. i. (State) the original source of energy for this food web.

Sunlight

··

no need for explanation

(1)

ii. (Name) an organism in this food web that is both a secondary and a tertiary consumer.

Adelie penguin

··

eats krill (secondary) and fish (tertiary)

(1)

b. Write in the names of organisms to form a complete food chain.

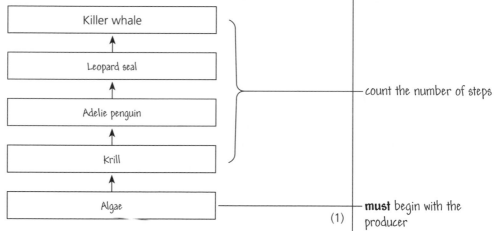

count the number of steps

must begin with the producer

(1)

c. There is concern that pollution of the environment may change the breeding grounds of the Adelie penguin.

(State and explain) the effect this might have on the populations of the Leopard seal and the Ross seal.

two pieces of information needed – can you link them with the word 'because'?

Leopard seal .....Population could fall because there would
························· be less food for them (so they would
························· breed less successfully)

Ross seal .....Population could rise because there would
························· be fewer leopard seals to act as predators

(4)

(Total 7 marks)

Cambridge IGCSE Biology 0610 Paper 2 Q6 June 2005 ▼

6. The graph below shows the heart rate and the cardiac output. The cardiac output is the volume of blood pumped out of the heart each minute.

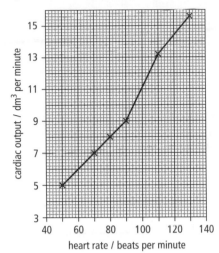

heart rate / beats per minute

a. i.   What is the cardiac output at a heart rate of 100 beats per minute? ———————
         11 dm³ per minute ——————

     read directly (and accurately!) from the graph

    (1)   don't forget the units!

    ii. Determine the increase in cardiac ouput when the heart rate increases from
         70 to 90 beats per minute.
              7⌐   ⌐9
         (9 – 7) = 2 .... dm³ per minute   (1)

         means the same as 'calculate' (or 'work out')

    iii. Determine the increase in cardiac output when the heart rate increases
         from 100 to 120 beats per minute.
              11⌐   ⌐14.4
         (14.4 – 11) = 3.4 .... dm³ per minute   (1)

b. i.   Which chamber of the heart pumps blood into the aorta? ——————————
         left ventricle ....

    (1)

    ii. The upper and lower chambers on each side of the heart are
         separated by valves.

         State the function of these valves.

         to prevent the back flow of blood (from ventricle to atrium) ....
         ....   (1)

    Simple recall of facts, you just have to know when to present them

    Cambridge IGCSE Biology 0610 Paper 2 Q7 November 2005

An experiment is designed to **test the validity of a hypothesis** and involves the **collection of data**.

e.g. light intensity affects the rate of photosynthesis

using appropriate apparatus and instruments

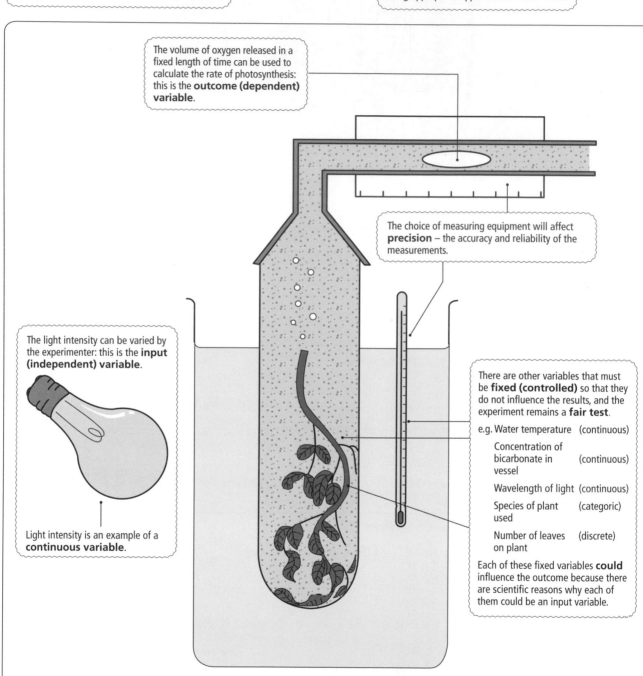

The volume of oxygen released in a fixed length of time can be used to calculate the rate of photosynthesis: this is the **outcome (dependent) variable**.

The choice of measuring equipment will affect **precision** – the accuracy and reliability of the measurements.

The light intensity can be varied by the experimenter: this is the **input (independent) variable**.

Light intensity is an example of a **continuous variable**.

There are other variables that must be **fixed (controlled)** so that they do not influence the results, and the experiment remains a **fair test**.

e.g. Water temperature   (continuous)

Concentration of bicarbonate in vessel   (continuous)

Wavelength of light   (continuous)

Species of plant used   (categoric)

Number of leaves on plant   (discrete)

Each of these fixed variables **could** influence the outcome because there are scientific reasons why each of them could be an input variable.

A '**control**' experiment is the same in every respect, **except** the input variable is not changed but is kept constant.

A control allows confirmation that **no unknown variable** is responsible for any observed changes in the responding variable; it helps to make the experiment a **fair test**.

A '**repeat**' is performed when the experimenter suspects that misleading data has been obtained through 'operator error'.

'**Means**' of a series of results minimize the influence of any single result, and therefore reduce the effect of any 'rogue' or anomalous data. The use of means improves the reliability of data – how confident you are about the observations or measurements.

May involve a number of steps

1. Organization of the **raw data** (the information which you actually collect during your investigation).

2. Manipulation of the data (converting your measurements into another form).

and

3. Representation of the data in **graphical** or other form.

Steps 1 and 2 usually involve **preparation of a table of results**

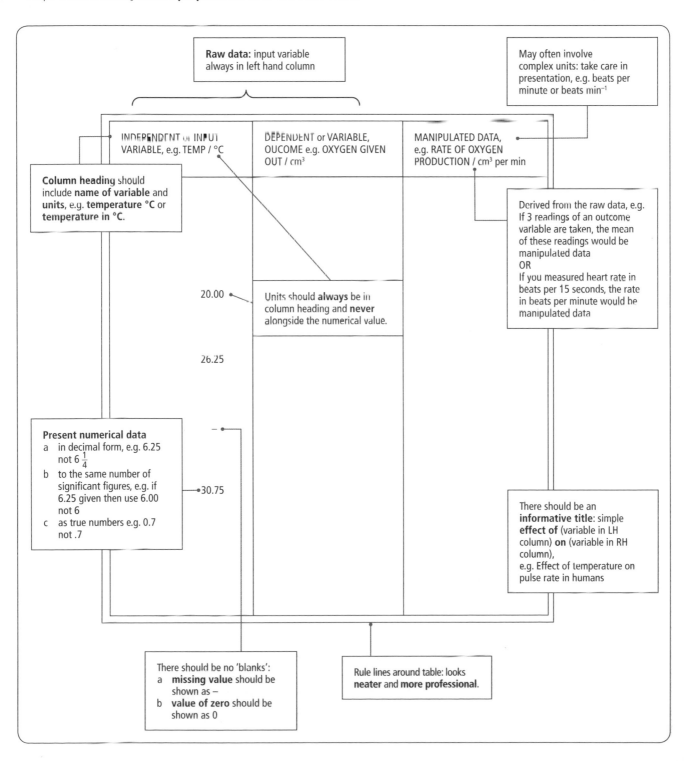

Raw data: input variable always in left hand column

May often involve complex units: take care in presentation, e.g. beats per minute or beats min⁻¹

INDEPENDENT or INPUT VARIABLE, e.g. TEMP / °C	DEPENDENT or VARIABLE, OUCOME e.g. OXYGEN GIVEN OUT / cm³	MANIPULATED DATA, e.g. RATE OF OXYGEN PRODUCTION / cm³ per min
20.00		
26.25		
30.75		

Column heading should include **name of variable** and **units**, e.g. **temperature °C** or **temperature in °C**.

Units should **always** be in column heading and **never** alongside the numerical value.

Derived from the raw data, e.g. If 3 readings of an outcome variable are taken, the mean of these readings would be manipulated data
OR
If you measured heart rate in beats per 15 seconds, the rate in beats per minute would be manipulated data

Present numerical data
a in decimal form, e.g. 6.25 not $6\frac{1}{4}$
b to the same number of significant figures, e.g. if 6.25 given then use 6.00 not 6
c as true numbers e.g. 0.7 not .7

There should be an **informative title**: simple **effect of** (variable in LH column) **on** (variable in RH column),
e.g. Effect of temperature on pulse rate in humans

There should be no 'blanks':
a **missing value** should be shown as –
b **value of zero** should be shown as 0

Rule lines around table: looks **neater** and **more professional**.

A graph is a visual presentation of data and may help to make the relationship between variables more obvious, i.e. allow the experimenter to **draw a conclusion**.

For example

AIR TEMPERATURE / °C	BODY TEMPERATURE OF REPTILE / °C
20	19.4
25	25.9
30	30.4
35	35.1
40	40.1
45	44.8

isn't as helpful as

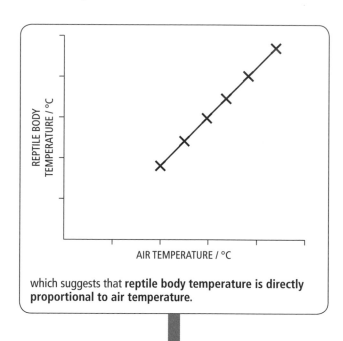

which suggests that **reptile body temperature is directly proportional to air temperature.**

Validity – how much confidence can be placed in the conclusion. Data is only valid if there is **only one input variable**. Can be affected by the **range** and **reliability** of measurements.

There should be an **informative title,** for example: **Effect of** (variable on *x* axis) **on** (variable on *y* axis).

A graph may be produced from a table of data **following certain rules**

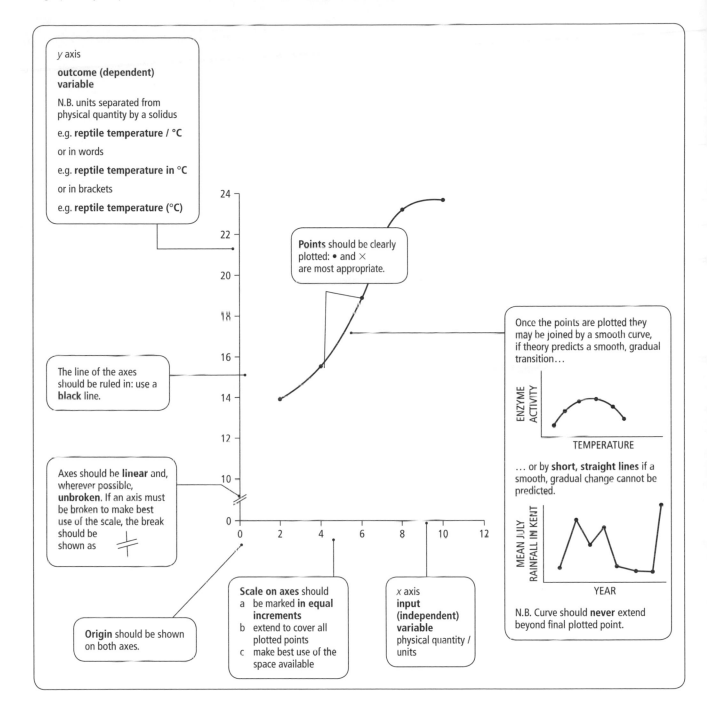

y axis

outcome (dependent) variable

N.B. units separated from physical quantity by a solidus

e.g. **reptile temperature / °C**

or in words

e.g. **reptile temperature in °C**

or in brackets

e.g. **reptile temperature (°C)**

Points should be clearly plotted: • and ×
are most appropriate.

The line of the axes should be ruled in: use a **black** line.

Once the points are plotted they may be joined by a smooth curve, if theory predicts a smooth, gradual transition…

ENZYME ACTIVITY

TEMPERATURE

… or by **short**, **straight lines** if a smooth, gradual change cannot be predicted.

MEAN JULY RAINFALL IN KENT

YEAR

N.B. Curve should **never** extend beyond final plotted point.

Axes should be **linear** and, wherever possible, **unbroken**. If an axis must be broken to make best use of the scale, the break should be shown as

Origin should be shown on both axes.

Scale on axes should
a   be marked **in equal increments**
b   extend to cover all plotted points
c   make best use of the space available

x axis

input (independent) variable

physical quantity / units

Chapter 6:
The variety of life

LIVING THINGS: THEIR CHARACTERISTICS

The biosphere is made up of living and non-living things. Plants and animals are living things because they have these characteristics:

RESPIRATION: the chemical reactions that break down nutrient molecules in living cells to release energy.
Food + oxygen ⟶ energy + water + carbon dioxide

Respiration is the most important process. No organism is alive unless it can release energy by respiration.

GROWTH: a permanent increase in size and dry mass by an increase in cell number or cell size or both.

MOVEMENT: an action by an organism or part of an organism causing a change of position or place.

EXCRETION: removal from organisms of toxic materials, the waste products of metabolism (chemical reactions in cells including respiration) and substances in excess of requirements.

NUTRITION: the process of taking in materials for energy, growth and development.
Plants require light, carbon dioxide, water and ions; animals need organic compounds, ions and water.

REPRODUCTION: the processes that make more of the same kind of organism.

SENSITIVITY: the ability to detect or sense changes in the environment (stimuli) and to make responses.

Key term: a **species** is a group of organisms that can reproduce to produce fertile offspring.

THE VARIETY OF LIVING ORGANISMS

A LIVING ORGANISM will show some or all of the characteristics of life – RESPIRATION is a particularly useful one to look for, as all living organisms need energy all of the time! You can find out which type of living organism it is by observing it carefully and then answering the following questions:

1	Is the organism made of a single cell?	YES	Go to question 2
		NO	Go to question 3
2	Do the cells have no clear nucleus?	YES	It's a **prokaryote**
		NO	It's a **protoctistan**
3	Do the cells have no cell wall?	YES	It's an **animal** – go to question 5
4	Do the cells have an obvious cell wall?	YES	Go to question 14
5	Does the animal have a backbone and internal skeleton?	YES	Go to question 6
		NO	Go to question 10
6	Does the animal have a smooth, moist skin?	YES	It's an **amphibian**
		NO	Go to question 7
7	Are there scales on the skin?	YES	Go to question 8
		NO	Go to question 9
8	Is the skin dry?	YES	It's a **reptile**
		NO	It's a **fish**
9	Does it have feathers, wings and a beak?	YES	It's a **bird**
		NO	It's a **mammal**
10	Does it have a segmented body?	YES	Go to question 11
		NO	It's a **mollusc**
11	Does it have jointed limbs?	YES	Go to question 12
		NO	It's an **annelid**
12	Does it have three pairs of legs?	YES	It's an **insect**
		NO	Go to question 13
13	Does it have four pairs of legs?	YES	It's a **spider**
		NO	It's a **crustacean**
14	Do the cells contain chlorophyll in chloroplasts?	YES	It's a **plant** – go to question 15
		NO	It's a **fungus**
15	Does it have stem, leaves and roots?	YES	Go to question 17
		NO	Go to question 16
16	Does it have stems and leaves but no roots?	YES	It's a **moss**
		NO	It's an **alga**
17	Does it produce spores?	YES	It's a **fern**
	Does it produce seeds?	YES	It's a **flowering plant (Angiosperm)**

NAMING THE ORGANISMS

Linnaeus was a scientist who believed he could put every organism into a group (the science of TAXONOMY) and give every organism a name (the science of NOMENCLATURE).

each organism has a unique, two-part name in his BINOMIAL SYSTEM e.g.

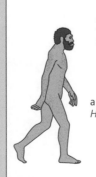

a human is *Homo sapiens*

a lion is *Panthera leo*

THE DIVERSITY OF LIFE

KINGDOM	GROUP	
Plant		
Flowering plants	Monocotyledons e.g. grasses	• only one 'seed leaf' • long thin leaves, with parallel veins
	Dicotyledons e.g. trees	• two 'seed leaves' • broad leaves with a network of veins
Ferns		• limited to damp environments because (a) they do not have a thick cuticle and (b) their gametes must swim through water during sexual reproduction • produce spores for wide dispersal
Animal		
Arthropods are invertebrates (animals without backbones). They have a hard exoskeleton and jointed limbs.	Insects	• three body parts – head, thorax and abdomen • three pairs of legs • usually one or two pairs of wings
	Crustaceans	• many segments, usually each with legs, claws or feelers • breathe via gills, as live in water • exoskeleton particularly hard, for protection
	Arachnids	• two body parts, eight legs and no wings • all have piercing jaws since all are predators
	Myriapods	• long, thin body with many segments for moving easily through soil and leaf litter • antennae as sense organs in dark habitats

HINT! It is easier to remember these groups if you use examples from near where you live!

VERTEBRATES (ANIMALS WITH BACKBONES)

CLASS	EXTERNAL FEATURES	OTHER FEATURES
Fish (all aquatic)	• Scales • Fins • Eyes and lateral line	• Jelly-covered eggs; usually use external fertilisation • Ectothermic • Gills for gas exchange
Amphibians (always breed in water)	• Moist skin • Four limbs • Eyes and ears	• Jelly-covered eggs; external fertilisation • Ectothermic • Lungs/skin for gas exchange
Reptiles (lay eggs on land)	• Dry, scaly skin • Four limbs (not in snakes) • Eyes and ears	• Soft-shelled eggs; internal fertilisation • Ectothermic • Lungs for gas exchange
Birds (very few are aquatic)	• Feathers (scales on legs) • Two wings, two legs • Eyes and ears	• Hard-shelled eggs; internal fertilisation • Endothermic • Lungs for gas exchange
Mammals (very few are aquatic)	• Fur or hair • Four limbs • Eyes and ears • Nipples	• Live young (a few lay eggs) • Endothermic • Lungs for gas exchange • Feed young with milk from mammary glands

You could be asked to directly describe these in exam questions

You could use these features in questions on other topics

Classification was originally based on **structure** (morphology and anatomy) but now the **sequence of bases in DNA** is used as a more accurate method. It is important to classify organisms for:
• conservation
• understanding of evolutionary relationships

1. Select from the list the name of the group of animals that best fits each description. Write your choice in the table.

arachnid   bird   crustacean   fish
insect   mammal   mollusc

DESCRIPTION OF ANIMAL	GROUP
a hard exoskeleton and more than 4 pairs of legs	
a hard shell and a slimy muscular foot	
one pair of wings and a beak	
one pair of wings and skin covered with fur	
two pairs of wings and one pair of antennae	

2. a. A biology student working in a museum mixed up a series of specimens. The specimens are named in this list:

AMPHIBIAN, EARTHWORM, BACTERIUM, BIRD, MAMMAL, FISH, REPTILE, FUNGUS, MOSS, FERN, PROTOCTISTAN, VIRUS, FLOWERING PLANT, INSECT, PARASITE

Choose organisms from this list to match the following descriptions:
   i.   body covered with fur; has mammary glands
   ii.   cells with cell wall but no chlorophyll; reproduce by spores
   iii. body covered with dry scales
   iv.   body has three regions – head, thorax and abdomen
   v.   single cell, without definite nucleus
   vi.   can only reproduce within another living cell; few genes inside protein coat
   vii. has feathers and a beak; maintains constant body temperature   (7)
   b. All living organisms display certain characteristics. Complete the following sentences about these characteristics.
   i.   releasing energy in cells is called…   (1)
   ii.   the removal of the waste products of the chemical processes in the body is called… (1)
   iii. the ability to respond to changes in the environment is called…   (1)

3. a. Diagrams A to H show organisms or parts of organisms (not drawn to scale).

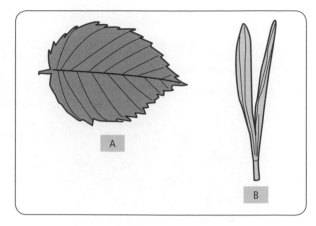

i.   State which of the drawings shows a monocotyledon leaf. State **one** reason for your choice.   (1)

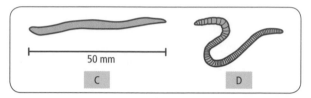

ii.   State which of the drawings shows an annelid. State **one** reason for your choice.   (1)

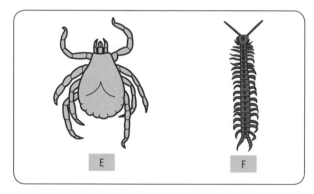

iii. State which of the drawings shows an arachnid. State **one** reason for your choice.   (1)

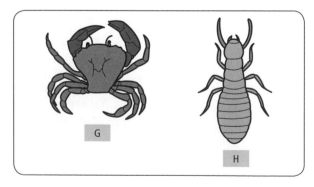

iv.   State which of the drawings shows a crustacean. State **one** reason for your choice. (1)
   b. The length of the drawing of worm **C** is shown. The actual length of the worm is 5 mm. Calculate the magnification of this drawing. Show your working.   (2)

Cambridge IGCSE Biology 0610 Paper 2 Q1 June 2007

4. The diagram below shows a mayfly nymph (a larva) that lives in water.

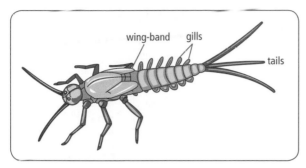

a. i. List two features, visible in the diagram, that show this is an insect. (2)
   ii. What special adaptation does the insect shown above have that allows it to live in water? (1)
b. Below there are five mayfly nymphs.

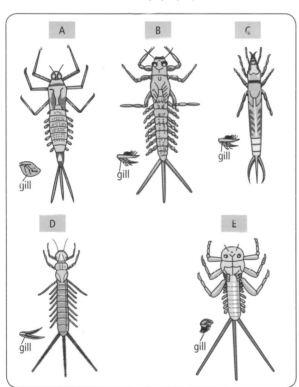

Use the key to identify the species of each mayfly.

		SPECIES
1	Rear pair of legs point towards tails ---------------------- go to 2	
	Rear pair of legs point forwards or sideways ---------- go to 3	
2	Gills project sideways from body	*Paraleptophlebia*
	Gills folded over body	*Ephemera*
3	Each gill a single flat plate ----------------------------------- go to 4	
	Each gill divided into two strands	*Potomanthus*
4	Tails 'feather' like in shape	*Centroptilum*
	Tails 'needle' shaped	*Ecdyonurus*

Write the diagram letter of each of the species in the correct box of the table.

SPECIES	DIAGRAM LETTER
Centroptilum	
Ecdyonurus	
Ephemera	
Paraleptophlebia	
Potomanthus	

(4)

Cambridge IGCSE Biology 0610 Paper 2 Q1 June 2005

5. The diagram shows ten organisms, **A** to **J**.

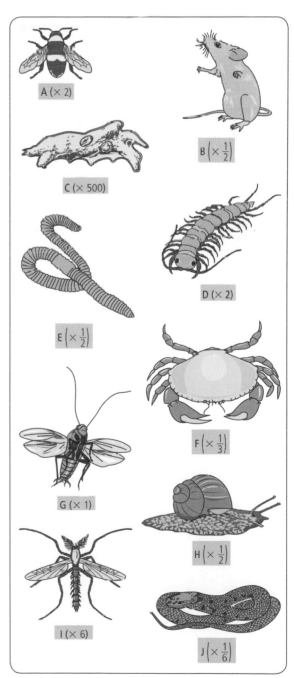

a. Look at the scale factor next to each organism. In real life,
which organism is the smallest?
which organism is the largest? (2)

b. **Five** of the organisms in the diagram have several features in common.
List the **letters** of these five organisms. (2)

c. i. Choose **three** of the organisms from your list in (b) which look to be more closely related to each other than they are to the other two organisms. (1)

ii. Give **two** features of these organisms which the other two organisms do **not** have. (2)

6. The drawings show four reptiles.

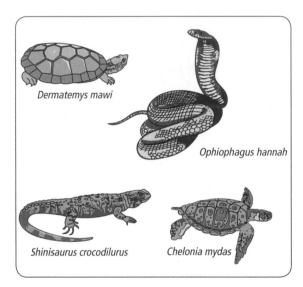

Dermatemys mawi

Ophiophagus hannah

Shinisaurus crocodilurus

Chelonia mydas

a. Complete the following key so that it can be used to identify each of the four reptiles. (4)

Key:

1 Has legs   go to 2
  No legs   ....
2 Most of body covered by a shell
                     Shinisaurus crocodilurus
3 Front legs   *Dermatemys mawi*
4 Front legs   *Chelonia mydas*

b. Describe **one** feature, **which you can see in the drawings**, which all four reptiles have in common but which is not present in a human. (1)

7. A large number of seeds were germinated on damp sand. Random samples of 10 seedlings were taken every two days.

The fresh mass and the dry mass of each sample were measured and are shown in the graph below.

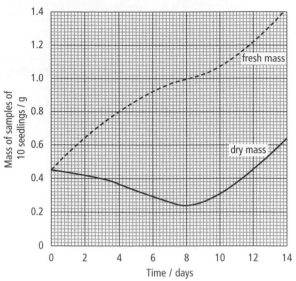

a. i. State why the fresh mass and dry mass of a seedling are different. (1)

ii. Fresh mass is not reliable as a measure of plant growth. Suggest why dry mass is a more reliable measure of plant growth. (1)

iii. Explain why 10 seedlings, rather than 1, were used for each sample. (1)

b. i. Describe what happens to the **fresh** mass of the seedlings in the first 2 days after the seeds were set to germinate. (2)

ii. Suggest a reason for this change in mass. (1)

c. i. Describe what happens to the **dry** mass of the seedlings during the first 8 days. (1)

ii. Suggest a reason for this change in mass. (2)

d. Suggest which processes begin in the living seed during the early stages of germination. (4)

LIVING ORGANISMS AND THEIR CHARACTERISTICS: Crossword

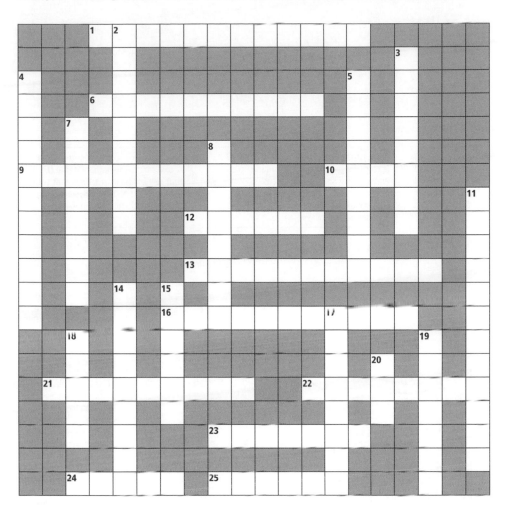

ACROSS:
1 The generation of new individuals of the same species
6 Single-celled organism with a definite nucleus
9 The ability to detect changes in the environment
10 A sequence of DNA – may be used in classification
12 This kingdom includes humans and other chordates
13 A change in the form or function of an organism
16 Oxidation of food to release energy
21 Measurable differences between organisms
22 A change in position of an organism
23 The peak of the hierarchy of taxonomic groups
24 Kingdom whose members feed by external digestion
25 Essential to drive the processes of life and to maintain organization

DOWN:
2 Removal of the waste products of metabolism
3 Genus that includes all the big cats
4 The 'proper' name for a human – means a 'wise man'
5 Single cell without a definite nucleus
7 The two-part system of naming living organisms
8 The father of naming organisms
11 The arrangement of molecules and cells into a working pattern
14 Supply of raw materials for metabolism, growth and reproduction
15 A permanent increase in size
17 Another name for 'classification' – the science of putting living organisms into groups
18 Kingdom with members capable of photosynthesis
19 Members of this taxonomic group can mate and produce fertile offspring
20 A set of questions that can be used to 'unlock' the name of an organism

Cells: building blocks of life

Living things are made of cells. Many of the chemical reactions that keep organisms alive (metabolic functions) take place in cells.

COMMON FEATURES OF CELLS

All cells (with a few exceptions) have these three things:

Cell membrane: a thin 'skin' that surrounds the cell contents. It controls the passage of dissolved substances into and out of the cell. This membrane is selectively permeable.

Cytoplasm: the contents of the cell (except for the nucleus). It is made up of water and dissolved substances. It also contains small structures **(organelles)** where chemical reactions take place.

Nucleus: the 'control centre' of the cell. It contains the genetic material **(DNA)** that carries the instructions that control the structure and activities of the cell. (Red blood cells do not have nuclei).

PLANT CELL FEATURES

Cell wall: a rigid (stiff) cell wall made of cellulose. This gives support. As a result, plant cells are fairly regular in shape. Water and dissolved substances can pass through the **permeable** cell wall.

Vacuole: the large, permanent vacuole contains water and dissolved substances **(cell sap)**. This helps to maintain pressure in the cells.

Chloroplasts: these contain **chlorophyll** and the **enzymes** needed for photosynthesis. They are found in the cells of green plants.

Stored food (starch): photosynthesis produces glucose (sugar). This is converted into **starch** and stored in the cytoplasm.

ANIMAL CELL FEATURES

Irregular shape: animal cells do not have a rigid cell wall so they are irregular in shape.

Denser cytoplasm: animal cells contain more dissolved substances and more organelles than plant cells. Animal cells contain more of the rod-like structures called **mitochondria** where aerobic respiration takes place. This is so they can release lots of energy quickly for fast movement. The cytoplasm also contains **ribosomes** on endoplasmic reticulum, where protein synthesis takes place.

Stored food (glycogen): carbohydrates are stored as glycogen in animal cells.

Vacuoles: animal cells may have several small temporary vacuoles. These can be for digestion or the excretion of excess water.

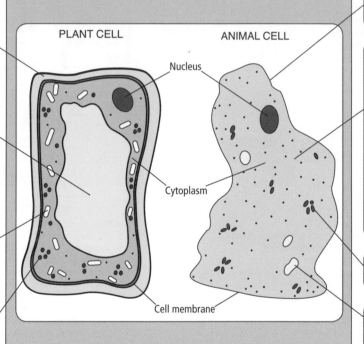

PLANT CELL   ANIMAL CELL

Nucleus

Cytoplasm

Cell membrane

Supplement

WORKING OUT THE SIZE OF CELLS: you will need to **measure** something and use a **magnification** to work out an **actual** size:

$$\text{MAgnification} = \frac{\text{Measured length}}{\text{Actual length}} \qquad \text{so} \qquad \text{Actual length} = \frac{\text{Measured length}}{\text{Magnification}}$$

Cells, tissues, organs and organ systems

Multicellular plants and animals contain many different types of cell. Each type of cell is designed for a particular function. Cells are organized to form tissues, organs, and organ systems. In a healthy **organism**, all the systems work together.

SPECIALIZED CELLS

A specialized cell is designed to do a particular job.
- Nerve cells have long fibres to carry messages.
- Muscle cells can contract and relax.
- Red blood cells carry oxygen around the body. They contain **haemoglobin**, which can combine with oxygen.

Specialized cells
- red cells carry oxygen
- white cells attack bacteria
- platelets help clotting

TISSUES

Large numbers of specialized cells working together make up **tissue**.

For example, blood tissue contains red cells for carrying oxygen, white cells for destroying harmful bacteria, and **platelets** to cause clotting in cuts.

Tissues
- blood vessels (capillaries)
- muscle layers (especially in the heart)
- blood

NB Arteries and veins are usually thought of as organs as they consist of several tissue layers.

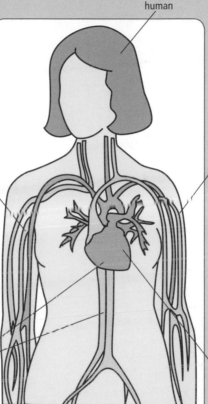

Organism
human

Organ system
blood circulation system

Organ
heart (to pump blood)

ORGANISM

Various organ systems together make up an **organism**.
You are a human organism. You have:
- a respiratory system
- a digestive system
- a circulatory system
- a nervous system
- an endocrine system

ORGAN SYSTEMS

Various organs together make up an **organ system**. For example, the **circulatory system** carries blood to all parts of the body. It is made up of the heart, the arteries, the veins, the capillaries and, of course, the blood.

ORGANS

Various tissues together make up an **organ**. Each organ has its own specific job. The heart, the stomach and the brain are all organs.

The **heart** has to pump blood around the body. It is made up of **muscle tissue**, **blood vessels** and **nerves**.

And for plants!

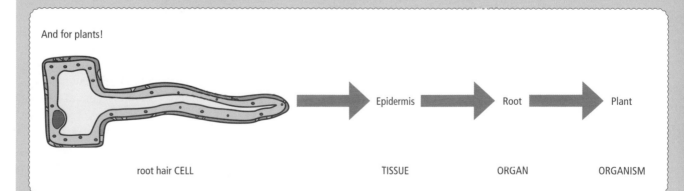

root hair CELL   TISSUE   ORGAN   ORGANISM

Epidermis → Root → Plant

1. The diagram below shows several types of animal and plant cell (not drawn to scale).

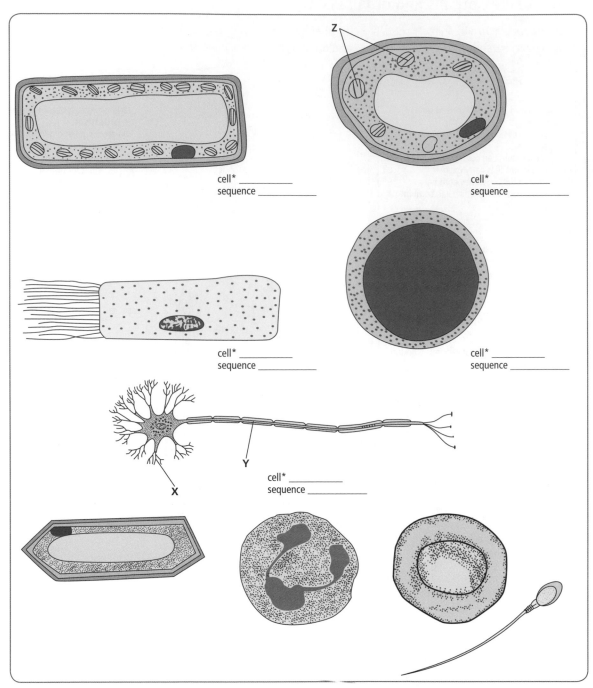

cell* _____
sequence _____

cell* _____
sequence _____

cell* _____
sequence _____

cell* _____
sequence _____

cell* _____
sequence _____

(5)

a. Use the key below to identify each of the cells marked with an asterisk (*). Write the letter corresponding to each cell next to the appropriate diagram. For each of the asterisked cells write the sequence of numbers from the key that led to your answer.

Identification Key

1 Cell with distinct cell wall .... go to 2
   Cell with membrane but no cell wall .... go to 4

2 Cell with chloroplasts in the cytoplasm .... go to 3
   Cell without chloroplasts in the cytoplasm ... CELL A

3 Cell with less than 10 chloroplasts visible ..... CELL B
   Cell with more than 10 chloroplasts visible .. CELL C

4 Cell containing a nucleus .... go to 5
   Cell lacking a nucleus .... CELL D

5 Cell with projections at one or more ends ... go to 6
   Cell without projections .... go to 8

6 Cell with projections at each end .... CELL E
   Cell with projection/projections at one end only ......
   .... go to 7

7 Cell with a large number of projections at one end ....
   .... CELL F
   Cell with a single, long projection at one end ....
   .... CELL G

8 Cell with a many-lobed nucleus .... CELL H
   Cell with a round nucleus .... CELL I

b. **Three** of the cells shown in the diagram above are types of human blood cell. Name each of the three cells and describe **one** function of each of the cells.

i.   Name of cell ....

Function ....

ii. Name of cell ....

Function ....

iii. Name of cell ....

Function .... (3)

2. The diagrams below show a plant cell, an animal cell and a virus. The diagrams are NOT drawn to the same scale.

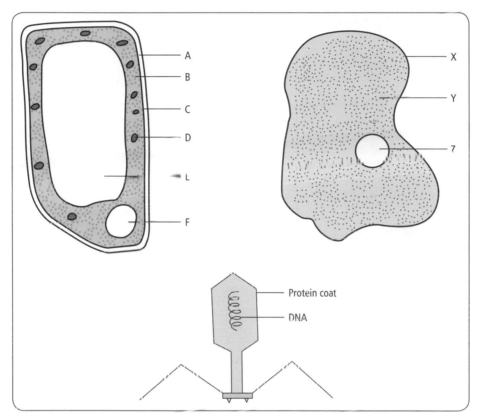

a. The parts of the plant cell are labelled A, B, C, D, E and F. Write the letter of the named part in the box next to its name in the list below.

CELL WALL ⎯

CYTOPLASM ⎯

VACUOLE ⎯ (3)

b. Name the part labelled **Z** in the diagram of an animal cell. (1)

c. State **two** differences, shown in the diagrams, between the plant cell and the animal cell. (2)

d. Which part of the plant cell contains the genes/alleles? (1)

e. What biological term describes a group of similar cells? (1)

f. Use **only** the information in the diagram to suggest **two** reasons why the virus is not a cell. (2)

3. The diagram shows a cell from a plant leaf.

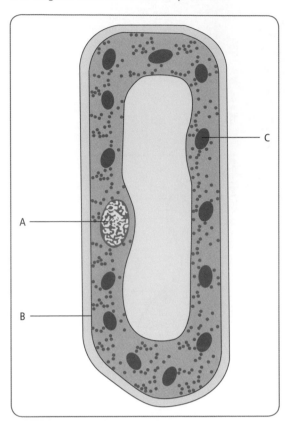

a. Name structures **A** and **B**. (2)

b. Structure **C** is a chloroplast. What is the function of a chloroplast? (1)

c. A plant cell has chloroplasts, whereas an animal cell does not. Give **two** more differences between plant and animal cells. (2)

d. An average plant cell is 50 μm long and 20 μm wide. How many plant cells could fit into 1 mm²? Show your working. (1)

4. a. The diagrams below show five types of tissue. Match each tissue with its correct function. (5)

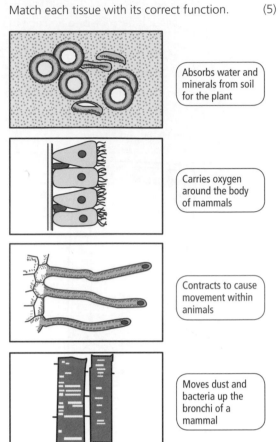

Absorbs water and minerals from soil for the plant

Carries oxygen around the body of mammals

Contracts to cause movement within animals

Moves dust and bacteria up the bronchi of a mammal

Transports water and minerals through the stem of a plant

b. Explain why the heart is described as an organ and not as a tissue. (2)

REVISION SUMMARY: Fill in the missing words

Use words from this list to complete the following paragraphs. The words may be used once, more than once or not at all.

PALISADE CELL, EPIDERMIS, TISSUES, EXCRETORY SYSTEM, SPECIALIZED, CELLS, BLOOD, KIDNEY, CHLOROPLASTS, LEAF, RED BLOOD CELL, DIVISION OF LABOUR, XYLEM, PHLOEM, NERVOUS, SYSTEMS, ENDOCRINE, ORGAN

a. Large numbers of ....that have the same structure and function are
    grouped together to form ...., for example .... Several
    separate tissues may be joined together to form an ...., which is a
    complex structure capable of performing a particular task with great efficiency. In
    the most highly developed organisms these complex structures may work together in
    ...., for example the .... in humans is responsible for the
    removal of the waste products of metabolism.   (6)

b. The structure of cells may be highly adapted to perform one function, i.e. the cells may
    become .... One excellent example is the .... which
    is highly adapted to carry oxygen in mammalian blood. If the different cells, tissues
    and organs of a multicellular organism perform different functions they are said to
    show .... One consequence of this is the need for close co-ordination
    between different organs – this function is performed by the .... and
    .... systems in mammals.   (5)

c. In plants an example of a cell highly specialized for photosynthesis is the
    ...., which contains many .... These cells are located
    in the organ called the ...., which also contains other tissues such as
    ...., which limits water loss, and ...., which transports
    water and mineral ions to the leaf.   (5)

Chapter 8:
Movement of molecules in and out of cells

Diffusion, osmosis and active transport

are processes by which molecules are moved. Diffusion and osmosis are passive, but active transport requires energy.

Supplement

DIFFUSION

The movement of molecules from a region of their higher concentration to a region of their lower concentration. Molecules move down a concentration gradient as a result of their random movement.
The energy for diffusion comes from the kinetic energy of random movement of molecules and ions.

DIFFUSION

This is a physical process that depends on the energy possessed by the molecules, so that
- small molecules diffuse faster than large molecules
- diffusion speeds up as temperature increases.

For living cells the principle (the movement of molecules down a concentration gradient) is the same, but there is one problem ⟶ the cell

is surrounded by a **cell membrane**, which can restrict the free movement of the molecules ⟶

Important examples:
- oxygen from air sacs in the lung to blood and from blood to cells
- soluble foods from gut to blood
- carbon dioxide from air to spaces inside leaf

This is a **partially permeable membrane**: the composition of the membrane (lipid and protein) allows some molecules to cross with ease, but others with difficulty or not at all. In this example the membrane is permeable to water ● but not to the larger sucrose molecule ◉ – the simplest sort of selection is based on the **size** of the molecules.

Diffusion is affected by
- surface area
- concentration gradient
- temperature
- distance

DIFFUSION OF WATER MOLECULES

FEWER WATER MOLECULES

MORE WATER MOLECULES

WATER CONCENTRATION GRADIENT

OSMOSIS IS THE DIFFUSION OF WATER

Water diffuses through partially permeable membranes by osmosis.

Osmosis causes water movement
- from tissue fluid to cells
- from soil water to root hairs
- from xylem to leaf mesophyll cells
Plants are supported by water, which has entered cells by osmosis, causing a pressure pushing against cell walls

Osmosis is
- the movement of water
- across a partially permeable membrane
- from a higher to a lower water potential
Water enters plant root cells by osmosis from the higher water potential in the soil. Water inside plant cells causes **turgor pressure** as it pushes against the inelastic cellulose cell wall.

Supplement

ACTIVE TRANSPORT MAY MOVE MOLECULES AGAINST A CONCENTRATION GRADIENT

ACTIVE TRANSPORT OF SOLUTE MOLECULES

MORE SOLUTE MOLECULES

FEWER SOLUTE MOLECULES

SOLUTE 'CONCENTRATION' GRADIENT

In this example there are more amino acid molecules ○ on the right side of the membrane than on the left – to move any more from left to right will be 'uphill', **against** the amino acid gradient. This active transport requires energy to 'drive' the molecules 'uphill' – this energy is supplied as adenosine triphosphate (ATP) from respiration.
Active transport is affected by any factor that affects respiration, e.g. temperature and oxygen concentration which is carried out by 'carrier proteins' ◗

Important examples are
- uptake of mineral ions from soil by root hair cells
- movement of sodium ions to set up nerve impulses

Supplement

---

1. The cells of the epidermis of the rhubarb stem contain a red pigment. This pigment makes it easy to see the cytoplasm in these cells, and to observe how the cytoplasm and cell membrane are affected by the external environment.

Strips of rhubarb epidermis were placed in three different sucrose solutions. The strips were left in the solutions for 30 minutes, and the cells were observed under the high power objective lens of a light microscope. The observations are recorded in the table opposite.

SOLUTION	APPEARANCE
A	Cell membrane pulled away from cell wall. No visible vacuole.
B	Cell membrane pushed firmly against cell wall. Large vacuole near centre of cell.
C	No change from 'normal' appearance – cell membrane just touching cell wall. Medium-sized vacuole in centre of cell.

a. Name the process responsible for these changes in the cells' appearance. (1)

b. What term is used to describe the cell membrane that allows this process to occur? (1)

c. What has happened to cause the cell membrane and cytoplasm to push against the cell wall in solution B? (1)

d. What can you say about solution C to explain the appearance of the cells? (1)

e. Copy this diagram. Complete it and add labels to show the cell membrane, cytoplasm and vacuole of a cell placed in solution A. (3)

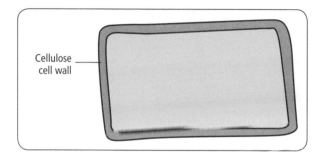

Cellulose cell wall

2. An experiment was set up to try to measure the concentration of the cytoplasm inside potato cells. A cork borer was used to produce identical cylinders of potato tissue. Twenty-one of these cylinders were blotted dry and weighed, and then three were placed into each of seven boiling tubes labelled A to G. The boiling tubes contained a series of sugar solutions, ranging from 0.0 to 0.6 molar. Two hours later the cylinders were removed from the tubes, blotted dry and reweighed. The results are shown in the table below.

TUBE	A	B	C	D	E	F	G
Concentration of sugar solution / molar	0.0	0.1	0.2	0.3	0.4	0.5	0.6
Percentage change in mass of potato cylinders	+10	+6	+2	−3	−7	−10	−12

a. Plot a graph of percentage change in mass against concentration of sucrose solution. (4)

b. Use your graph to state the concentration of sucrose solution that will give no change in mass of the potato tissue. (2)

c. Use your biological knowledge to explain how this value allows you to work out the concentration of the cytoplasm in the potato cells. (2)

d. Explain why **three** potato cylinders were placed into each of the solutions. (2)

e. Explain why the potato cylinders are blotted dry before they are weighed. (2)

f. Three more potato cylinders were placed into a 1 molar solution of sucrose. The potato cylinders weighed a total of 10 g. Predict the mass of this tissue after two hours in this solution. (2)

3. Crop production in many areas of the world needs the application of large volumes of water. However, when the water evaporates from the soil, traces of salts are left behind. After several years, the soil becomes too salty for most plants to grow in it.

a. i. State **three** functions of water in plants. (3)

ii. With reference to the water potential gradient, explain why plants may die when grown in salty soil. (3)

b. Some plants are able to pump salts out of their roots.

i. Name the process plants could use to pump salts out of their roots. (1)

ii. Suggest how the process named in **(i)** could affect the rate of growth of the plants if the process was operating all the time. (2)

iii. Plants need mineral salts for normal, healthy growth. Name **two** minerals that plants need and state their functions. (4)

Adapted from Cambridge IGCSE Biology 0610
Paper 3 Q5 June 2006

SUMMARY REVIEW: Fill in the missing words

Use words from this list to complete the following paragraphs. You may use each word once, more than once or not at all.

DIFFUSION, OSMOSIS, PHOTOSYNTHESIS, ACTIVE TRANSPORT, CELL WALL, SWELL, SHRINK, PARTIALLY PERMEABLE MEMBRANE, CYTOPLASM, RESPIRATION, OXYGEN, GLYCOGEN, CARBON DIOXIDE, AMINO ACIDS, PATHWAY, ENERGY, ALONG, AGAINST, CONCENTRATION GRADIENT, EPIDERMIS, LOWER

Animal cells contain ...., a semi-fluid solution of salts and other molecules, and are surrounded by a .... When surrounded by distilled water, the animal cells .... because the cell has a .... water potential than the surrounding water. Plant cells do not have this problem because they are surrounded by a ....

In the gut soluble food substances such as .... cross the gut lining into the capillaries by the process of ...., which is the movement of molecules down a .... When an equilibrium is reached between the gut contents and the blood, glucose may continue to be moved using the process called ...., which consumes .... and can move molecules .... a concentration gradient.

The leaves of green plants obtain the gas ...., which they require for the process of photosynthesis, by the process of .... They also lose the gas oxygen, produced during .... by the same process. (14)

Chapter 9:
Biological molecules and food tests

WHAT WOULD YOU TEST?

Examples include
- urine for glucose and protein
- factory outflow for waste starch
- spinal fluid for glucose (concentration falls if patient has meningitis)
- food for diabetic people for glucose

Biological molecules all contain carbon, hydrogen and oxygen.

Proteins also contain nitrogen and sometimes sulfur.

MOLECULE	CARBON (C)	HYDROGEN (H)	OXYGEN (O)	NITROGEN (N)
Lipid	√	√	√	X
Carbohydrate	√	√	√	X
Protein	√	√	√	√

Large molecules are usually made up of similar smaller molecules which can be broken down by hydrolysis, which uses water, and joined together in new combinations through condensation. For example, starch and glycogen are both combinations of glucose.

LIPIDS (FATS and OILS) (contain the elements C, H and O)
- gently shake the mixture
- pour the lipid/alcohol mixture into an equal volume of water
- a MILKY-WHITE EMULSION is a positive result: lipid is present

MIX A SAMPLE WITH ETHANOL

CONTROLS
- pure water should give a negative result: test solutions are not contaminated
- a known solution of the food should give a positive result: the test solutions are working properly

TESTING FOR CHEMICALS

DISSOLVE A SAMPLE IN WATER

VITAMIN C
- add drops of the solution to the blue dye DCPIP until the blue colour disappears
- few drops - strong vitamin C solution
- many drops - weak vitamin C solution

PROTEINS (contain the elements C, H, O and N)
- add a few drops of Biuret reagent
- gently shake the mixture
- a PURPLE/MAUVE COLOUR is a positive result: protein is present

GLUCOSE (a reducing sugar)
- add 2 cm³ of Benedict's Reagent
- heat the mixture for 2–3 minutes in a boiling water bath
- a BRICK-RED/ORANGE COLOUR is a positive result: glucose is present

GLUCOZA

Glucose and starch are both CARBOHYDRATES and contain the elements C, H and O.

STARCH
- add a few drops of iodine solution
- gently shake the mixture
- a BLUE-BLACK COLOUR is a positive result: starch is present

---

1. Food contains biological molecules. It is sometimes important to know exactly which molecules are present in which foods. There are chemical tests called FOOD TESTS that allow us to find this out.
The boxes opposite list certain molecules and food tests.

a. The molecules and the chemical tests are only correctly matched in four of the boxes. Write down the letters of these four boxes. (4)

b. Describe exactly how you would carry out a Benedict's test if you were given a powdered sample of a food. Be sure to include any safety measures you would take. (3)

A	Protein	Biuret reagent
B	Simple sugar	Iodine solution
C	Fat	Biuret reagent
D	Starch	Benedict's reagent
E	Protein	Alcohol emulsion
F	Starch	Iodine solution
G	Fat	Alcohol emulsion
H	Simple sugar	Benedict's reagent

2. The figure below shows how four different food tests (**1**, **2**, **3** and **4**) were carried out. Each gave a positive result.

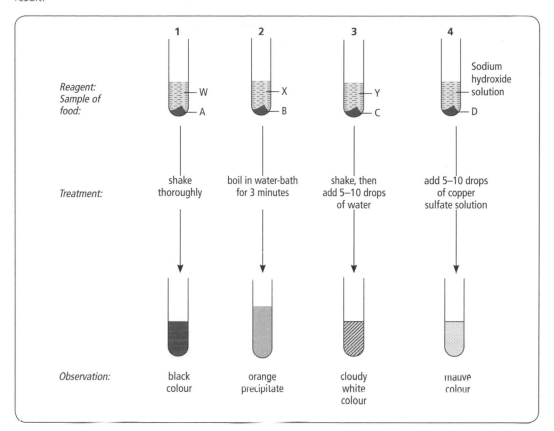

	1	2	3	4
Reagent: Sample of food:	W A	X B	Y C	Sodium hydroxide solution D
Treatment:	shake thoroughly	boil in water-bath for 3 minutes	shake, then add 5–10 drops of water	add 5–10 drops of copper sulfate solution
Observation:	black colour	orange precipitate	cloudy white colour	mauve colour

a. Complete the table giving the names of the reagents **W**, **X** and **Y**, and the type of nutrient shown to be present by each of the four tests.

	TEST			
	1	2	3	4
Reagent				
Nutrient in sample of food				

(5)

b. State the colours you would have seen if the results were negative in:

test 1 ....

test 2 ....

test 4 .... (3)

c. i   Suggest **two** possible items of a person's diet which are rich in the nutrient in the food **C**.

ii   Suggest **two** possible items of a person's diet which are rich in the nutrient in the food **D**.

(2)

3. It is possible to carry out simple chemical tests to investigate the chemical content of biological solutions. Usually these tests involve mixing reagents with the solution, and noting any colour change. The following are the results obtained from a series of these tests.

SOLUTION	COLOUR AFTER TESTING WITH REAGENT			
	IODINE SOLUTION	BENEDICT'S REAGENT	BIURET REAGENT	BENEDICT'S AFTER ACIDIFICATION AND NEUTRALIZATION
A	Blue-black	Clear blue	Clear blue	Clear blue
B	Straw yellow	Orange	Purple	Orange
C	Straw yellow	Clear blue	Clear blue	Orange
D	Blue-black	Clear blue	Faint purple	Clear blue
E	Straw yellow	Orange	Clear blue	Orange

Based on the results, match A-E in the table with the solutions listed below.

a. The waste water from a laundry   (1)

b. Milk   (1)

c. Crushed potato   (1)

d. Urine from a sufferer from sugar diabetes   (1)

e. Sweetened tea   (1)

4. The bar chart below shows the percentage of each of the main elements in the human body.

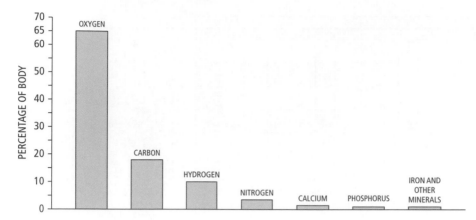

a. Convert this information
   i. into a table
   ii. into a pie chart (5)
b. Which do you think is the best way of showing the data? Explain why. (2)
c. These elements are largely present as molecules. The proportions of the main molecules in a human body are approximately:

   Fats   14%
   Proteins   12%
   Carbohydrates   1.0%
   Water   70%

   i. Which is the most abundant molecule? (1)
   ii. Which molecule contains most of the nitrogen? (1)
   iii. Which molecule contains most of the oxygen? (1)
   iv. Which molecule contains most of the carbon? (1)
   v. The total of these molecules does not add up to 100 per cent. Suggest another organic compound that forms a proportion of the remainder. (1)
   vi. Which structure(s) in the body will contain most of the calcium? (1)

5. Simple chemical tests on foods can show which type of nutrients they contain.

A group of students was given samples of four different foods to test – the samples were prepared by dissolving powdered food in distilled water. The four foods were identified as A, B, C and D. They were also given a sample of distilled water.

They carried out three tests on the five samples. Test 1 was for glucose, test 2 was for protein and test 3 was for starch. The results of the tests are shown in the table.

FOOD TEST	1 GLUCOSE	2 PROTEIN	3 STARCH
Sample A	Blue	Purple	Brown
Sample B	Orange	Blue	Brown
Sample C	Blue	Blue	Black
Sample D	Orange	Purple	Brown
Distilled water	Blue	Blue	Brown

a. Which colour indicates the presence of starch? (1)
b. What is the name of the solution which gives a purple colour if protein is present? (1)
c. Which one of the samples contained protein and glucose but no starch? (1)
d. Why were the tests also carried out on distilled water? (1)
e. This series of tests does not detect the presence of fat. Describe a test for fat, including a positive result for the test. (3)
f. Explain **one** danger to health of eating too many foods which contain large amounts of fat. (1)

TESTING FOR CHEMICALS: Crossword

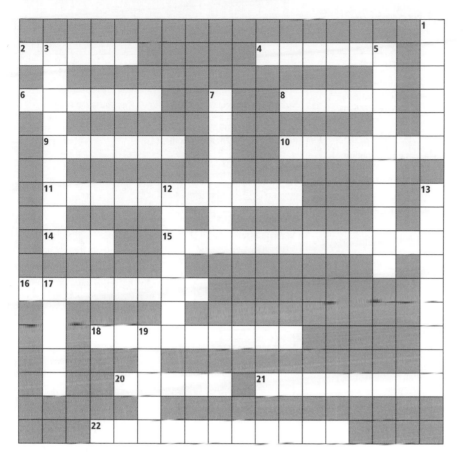

ACROSS:

2 The most common biological molecule – all biochemical reactions take place dissolved in it
4 Important storage carbohydrate in plants
6 Reagent used to test for protein
8 Biological fluid sampled by doctors testing for anaemia
9 Colour of a positive Benedict's test
10 The most common carbohydrate used for respiration
11 Steroid that can be harmful to blood vessels
14 Abbreviation for the genetic material in plants and animals
15 Reaction in which two molecules are joined together with the elimination of water
16 A test for lipids – millions of tiny fat globules suspended in water
18 Colour of a positive starch test
20 Group of molecules including fats and oils
21 Vital for life, but no single test for them
22 The most common protein in 8 across

DOWN:

1 Colour of positive test for protein
3 Subunit of a protein (5,4)
5 Reaction in which a large molecule is broken down to smaller ones by the addition of the elements of water
7 Use the Biuret test to check if it's present
12 The sweetest carbohydrate
13 Reagent that turns orange when heated with 10 across
17 Appearance of the mixture of lipid, ethanol and water
19 The easiest human body fluid to obtain for medical testing

Chapter 10:
Enzymes control biological processes

Enzymes control biological processes
and are widely exploited by humans

Enzymes
- are proteins that function as biological catalysts (a catalyst increases the rate of a chemical reaction but is not changed by the reaction)
- are important in living organisms because they make reactions occur fast enough to sustain life.

The **shape** of the active site enables the enzyme to 'recognize' its substrate **in a very specific way**. Any factor that alters the enzyme's shape will affect its activity. Important examples are changes in temperature and pH.

ENZYMES ARE PROTEINS WHICH ACT AS BIOLOGICAL CATALYSTS

Substrate molecules fit exactly onto an ACTIVE SITE on the enzyme to form an ENZYME-SUBSTRATE COMPLEX: this is the LOCK AND KEY hypothesis

Substrate molecules react together to form a **product** which leaves the active site

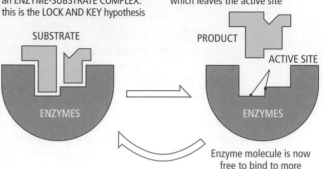

Enzyme molecule is now free to bind to more substrate molecules

Supplement

pH: is a measure of acidity or alkalinity, and is a mathematical method for expressing the concentration of H^+ ions in solution.

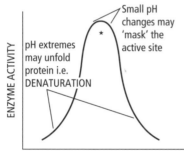

pH extremes may unfold protein i.e. DENATURATION

Small pH changes may 'mask' the active site

* The optimum pH for an enzyme depends on its site of action, pH enzymes in the stomach (where HCl is present) have an optimum about pH 2 but intestinal enzymes (no HCl) have optimum about pH 7.5.

TEMPERATURE: like all proteins, enzymes are made up of long, precisely folded chains of amino acids. This folding may be 'undone' by high temperature so that the enzyme may lose its active site – it is **denatured**.

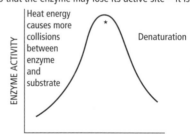

Heat energy causes more collisions between enzyme and substrate

Denaturation

* The optimum temperature for human enzymes is close to 37 °C. For most plants it is lower.

The activity of enzymes can be investigated. For one example see question 2 on p. 11

Enzymes may be intracellular or extracellular

These are both made **and** have their action inside cells ('intra' means 'inside')
e.g. photosynthetic enzymes inside chloroplasts or respiratory enzymes inside mitochondria.

These are **made** inside cells but **have their action** outside the cell ('extra' means 'outside')
e.g. digestive systems in the human gut or enzymes released by saprotrophic fungi and bacteria.

THE SPECIFICITY AND CATALYTIC ACTIVITY OF ENZYMES MAKE THEM VERY USEFUL TO HUMANS

ENZYMES HELP SEED GERMINATION
- **Amylase** breaks down starch to sugars
- **Lipase** breaks down fats (e.g. in sunflower seeds) to fatty acids and glycerol.

GENETIC ENGINEERING
- **Restriction enzymes** are used to cut out specific genes, and to open up bacterial plasmids
- **Ligases** are used to 'stitch' human genes into bacterial plasmids.

COMMERCIAL
- **Proteases** help to soften leather for the garment industry
- **Lipase** removes stains from clothing – component of 'biological' washing powders
- **Lactase** breaks down lactose to provide lactose-free milk
- **Pectinase** breaks down small pieces of plant tissue to turn cloudy fruit juice into clear fruit juice.

1. Use words from this list to complete the following paragraph about enzymes. You may use each word once, more than once or not at all.

PATHWAYS, ENZYMES, CATALYSTS, ACTIVATORS, PROTEINS, UNUSUAL, SPECIFIC, DENATURED, DESTROYED

Enzymes are .... that speed up the biochemical .... in living organisms. The enzymes themselves are not changed in these reactions, that is they are biological ....
       Enzymes are ....– each of them controls only one type of reaction. They are ....by high temperatures and by extremes of pH. (5)

2. In an experiment, apparatus is used to measure the effect of one factor, the **input** or **independent** variable, on the value of a second factor, the **outcome** or **dependent** variable. To be sure that the experiment is a fair test, all other factors must be kept constant – these are **fixed** variables. For example, a simple closed manometer may be used to measure the effect of temperature on the activity of the enzyme catalase.

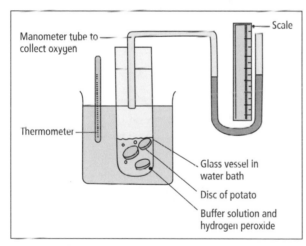

Manometer tube to collect oxygen
Scale
Thermometer
Glass vessel in water bath
Disc of potato
Buffer solution and hydrogen peroxide

a. What is the input (independent) variable in this experiment?
b. Suggest two fixed variables, and explain how the experimenter could keep them constant.
c. What is the outcome variable in this experiment? How could it be measured?
d. Suggest a suitable control for this experiment.
e. Students using this apparatus collected five sets of data, and used them to calculate a **mean**. How does this improve the **validity** of the results? (5)

3. Respiration is the process that releases energy in cells. It is a process that depends on enzymes. A group of students were interested in the effects of temperature on the rate of oxygen consumption by maggots. They obtained the following results.

TEMPERATURE / °C	OXYGEN CONSUMPTION / mm per second
15	0.3
25	0.6
35	1.1
45	0.8
55	0.2
65	0.0

a. Plot these results in the form of a graph, and explain the shape of the curve. (6)
b. In this investigation, identify the **input** variable and the **outcome** variable. Suggest any **fixed** variables. (4)

4. a. Define the term *enzyme*. (2)
   b. Enzymes are used in biological washing powders.
      i. Describe how the presence of these enzymes may increase the efficiency of the washing powder in removing stains from clothes. (3)
      ii. Explain why the temperature of the wash needs to be carefully controlled. (3)
      iii. Suggest a suitable temperature for a wash using a biological washing powder. Explain your answer.
         Suitable temperature ....
         Explanation .... (1)
   c. Outline how enzymes can be manufactured for use in biological washing powders. (4)

Cambridge IGCSE Biology 0610 Paper 3 Q6 November 2005

5. Catalase, an enzyme, is present in all living cells including those of potato and liver. It speeds up the breakdown of hydrogen peroxide as shown by the equation:

$$\text{hydrogen peroxide} \xrightarrow{\text{catalase}} \text{oxygen} + \text{water}$$

The oxygen is given off as a gas which can be collected over water, as shown below.

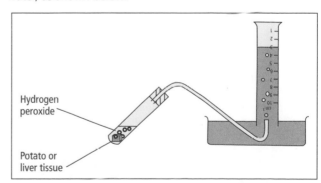

Hydrogen peroxide
Potato or liver tissue

Two different tissues, potato and liver, were used for this investigation. Samples, each of one gram, were prepared from both tissues. Some of the samples were left raw and others were boiled. Some samples were left as one cube and others were chopped into small pieces as shown in the first table below. 2 cm³ hydrogen peroxide was added to each sample. The volume of oxygen produced in five minutes was collected in the measuring cylinders, as shown in the table.

SAMPLE	A	B	C	D
Treatment	raw	raw	boiled	boiled
Results for potato				
Results for liver				

a. i. Complete the following table by reading the values for oxygen collected in the measuring cylinders in the table above

TISSUE	VOLUME OF OXYGEN COLLECTED FROM EACH SAMPLE / cm³			
	A	B	C	D
Potato				
Liver				

(2)

ii. Plot the volumes of oxygen collected from the samples as a bar chart.

iii. Describe the difference in results between sample **A** for potato and sample **A** for liver. (2)

iv. There is a difference between the samples for **A** and **B** for liver. Suggest an explanation for this difference. (2)

b. State the importance of samples **C** and **D** in this investigation. (1)

c. Suggest how you could test that the gas given off was oxygen. (1)

Adapted from Cambridge IGCSE Biology
0610 Paper 6 Q1 June 2005

Supplement

6. The diagram shows the main features of a biosensor used to monitor blood glucose levels.

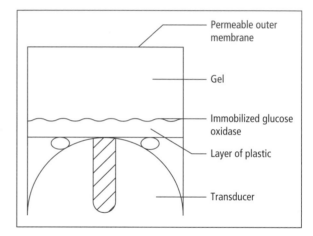

Permeable outer membrane

Gel

Immobilized glucose oxidase

Layer of plastic

Transducer

a. The biosensor uses the enzyme, glucose oxidase, which is immobilized. Explain what is meant by *immobilized*. (1)

b. Glucose oxidase catalyses the reaction:

Glucose + Oxygen ⟶ Gluconic acid + Hydrogen peroxide

Glucose diffuses into the biosensor through the outer membrane and gel. Explain how the oxygen concentration around the transducer will be affected. (2)

c. What is the function of the transducer? (1)

ENZYMES CONTROL BIOLOGICAL PROCESSES: Crossword

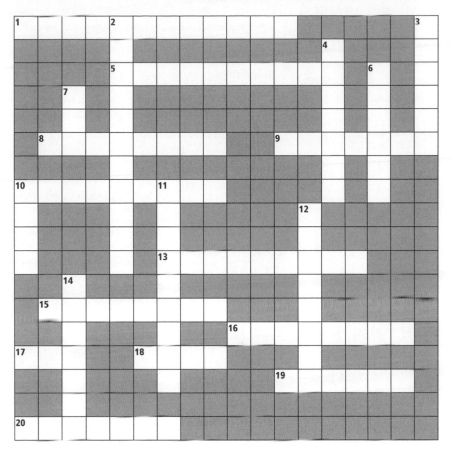

ACROSS:

1 The loss of an enzyme's three-dimensional shape
5 All of the chemical reactions occurring in cells
8 More of these ions lower the pH
9 A poison that affects an enzyme in respiration
10 A reactant in an enzyme-catalysed reaction
13 A compound that reduces the activity of an enzyme
15 A protein-digesting enzyme
16 A molecule that speeds up a reaction but is not affected itself
17 With 18 across and 7 down – a mechanism to explain enzyme action
18 See 17 across
19 A stomach enzyme with a very low optimum pH
20 The 'best' value for a factor affecting enzyme activity

DOWN:

2 This factor affects an enzyme's activity – too high and 1 across may be the result
3 Fat-digesting enzyme – useful in biological washing powders
4 Starch-digesting enzyme
6 With 10 down – the part of an enzyme where the reaction occurs
7 See 17 across
10 See 6 down
11 A molecule that can speed up an enzyme-catalysed reaction
12 Enzyme that breaks down harmful ions in liver cells
14 Molecule produced in an enzyme-catalysed reaction

An ideal human diet

contains fat, protein, carbohydrate, vitamins, minerals, water and fibre **in the correct proportions**. For a list of the component atoms of each type of molecule, see Chapter 9 (p. 30).

An adequate diet provides sufficient **energy** for the performance of metabolic work, although the 'energy' could be in any form.

A balanced diet provides all dietary requirements **in the correct proportions**. Ideally this would be $\frac{1}{7}$ **fat**, $\frac{1}{7}$ **protein** and $\frac{5}{7}$ **carbohydrate**.

In conditions of **under**nutrition the first concern is usually provision of an **adequate diet**, but to avoid symptoms of **mal**nutrition a **balanced diet** must be provided.

CARBOHYDRATES

These are principally an energy source. A **respiratory substrate** is oxidized to release **energy** for active transport, synthesis of macromolecules, cell division and muscle contraction.

Common sources: rice, potatoes, wheat and other cereal grains, i.e. as **starch**, and as refined sugar, **sucrose**, in food sweetenings and preservatives.

FLOUR

Digested in mouth and small intestine and absorbed as **glucose**.

LIPIDS

These are an energy source (they are highly reduced and therefore can be oxidized to release **energy**). Also important in **cell membranes** and as a component of **steroid hormones**.

Common sources: meat and animal foods – rich in **saturated fats** and **cholesterol**, plant sources such as sunflower and soya – rich in **unsaturated fats**.

Digested in the small intestine and absorbed as **fatty acids and glycerol**.

BUTTER MILK

VITAMINS

VITAMINS have no common structure or function, but are essential in small amounts to use other dietary components efficiently. Their absence results in **deficiency diseases**.

	Source	Deficiency causes
e.g. **vitamin C** (water soluble)	Citrus fruits	**Scurvy** – bleeding gums, slow healing of wounds
vitamin D (fat soluble)	Cod liver oil, margarine	**Rickets** – misshapen and poorly growing bones

Supplement

PROTEINS

Are **building blocks** for growth and repair of many body tissues (e.g. myosin in muscle, collagen in connective tissues), as **enzymes**, as **transport systems** (e.g. haemoglobin), as **hormones** (e.g. insulin) and as **antibodies**.

Common source: meat, fish, eggs and legumes/pulses. Must contain eight **essential amino acids** since humans are not able to synthesize them. (Animal sources generally contain more of the essential amino acids than vegetable sources.) Digested in stomach and absorbed as **amino acids**.

BAKED BEANS

Deficiency of protein causes poor growth – in extreme cases (in less economically developed countries) may cause **marasmus** or **kwashiorkor**. Usually energy foods are also in short supply. This is **protein-energy malnutrition**.

Supplement

WATER is required as a solvent, a transport medium, a substrate in digestive reactions and for lubrication (e.g. in tears). A human requires 2–3 dm³ of water daily – most commonly from drinks and liquid foods.

MINERALS have a range of **specific** roles, and absence may cause **deficiency diseases**
Deficiency causes

	Source	Deficiency causes
e.g. **Calcium** (Ca^{2+})	Dairy products	Poor growth of bones and teeth
Iron (Fe^{2+})	Red meat, spinach	Anaemia – poor oxygen transport in blood

They are usually ingested with other foods, but supplements may be necessary (e.g. iron tablets are sometimes needed following menstruation).

Supplement

FIBRE (originally known as **roughage**) is mainly cellulose from plant cell walls and is common in fresh vegetables and cereals. It **may** provide some energy but mainly serves to aid the formation of faeces and prevent constipation.

DANGER!

Some foods contain additives, e.g. PRESERVATIVES (such as nitrates) to slow down spoilage of food and COLOURINGS to make food attractive. Excess consumption of additives can be dangerous, for example some food colourings can cause **hyperactivity**.

Supplement

THE IDEAL DIET may vary according to

- **Age**: young people need extra protein for growth and carbohydrates for energy
- **Gender**: boys use more 'energy foods' as they are heavier and often more active; girls need more iron to replace blood lost during menstruation
- **Activity**: more exercise, including manual work, increases the need for energy sources such as fats and carbohydrates.

REPRODUCTION IS ALSO SIGNIFICANT

- **Pregnancy**: women need more protein for general growth of the fetus, and both calcium (for bones and teeth) and iron (for production of blood)
- **Breastfeeding**: production of milk means that the mother needs extra energy, protein and the same minerals as during pregnancy

MALNUTRITION is the result of an unbalanced diet

starvation	a shortage of fats/carbohydrates means the body has too little energy. Protein may be broken down to provide energy, causing wasting of muscle
constipation	too little dietary fibre means that peristalsis does not occur and faeces are not expelled from the intestines
coronary heart disease	too much fat, cholesterol and salt increase the risk of CHD
obesity	too much fat or carbohydrate may mean energy intake exceeds energy need – the excess is stored and body mass increases
scurvy	a shortage of vitamin C means that connective tissues are not formed – teeth may fall out and wounds do not heal properly

FOOD SUPPLY AND FAMINE

DAMAGE BY PESTS
- especially in monocultures
- pesticides may be overused so resistant pests evolve

ABSENCE OF WATER
- global warming may affect rainfall patterns
- deforestation can affect the water cycle
- diversion of water supplies, e.g. for hydroelectric power schemes

DESERTIFICATION
- soil erosion
- deforestation
- overgrazing

FLOODING
- fertile soil can be eroded
- plant roots can be deprived of oxygen
- crop plants can be damaged

COST OF FUEL AND FERTILISER
- machinery may be too expensive to run
- yield may be very low if fertilizer cannot be used

WAR
- crops may be burned to deprive people of food
- farm workers may be forced to join armies
- working on farms may become too dangerous

Oil

COST OF TRANSPORTATION
- food surpluses cannot be equally distributed

URBANIZATION
- buildings take up growing space
- cities can affect local temperatures

INCREASING POPULATION
- may be too many people for available food
- domestic animals may graze on land previously used for crops
- next year's seeds may be eaten as food supplies run out

1. The table shows the composition and energy content of four common foods.

FOOD	ENERGY CONTENT / kJ g⁻¹	COMPOSITION PER 100g					
		Protein / g	Fat / g	Carbohydrate / g	Vitamin C / mg	Vitamin D / mg	Iron / mg
A	3700	0.5	80	0	0	40	0
B	150	1.2	0.6	7	200	0	0
C	400	2.0	0.2	25	10	0	8
D	1200	9.0	1.5	60	0	0	0

a. Which food would be best to prevent rickets?
b. Which food would be best for a young man training for cross-country running?
c. Which food would be most needed by a menstruating woman?
d. Which food would be the most useful to a body-builder?
e. Which food would be most dangerous for a person with heart disease? (5)

2. Orange juice is a good source of vitamin C. Three different brands of orange juice were bought from a health store, and another sample was prepared by squeezing fresh oranges. Each of the four juices was tested as shown in the diagram below. A solution containing 0.1 per cent ascorbic acid (vitamin C) was tested at the same time.

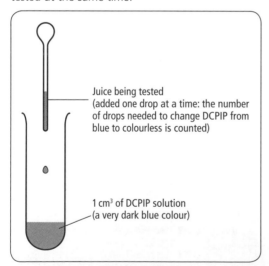

Juice being tested (added one drop at a time: the number of drops needed to change DCPIP from blue to colourless is counted)

1 cm³ of DCPIP solution (a very dark blue colour)

The results of this test are shown in the table below.

JUICE TESTED	VOLUME NEEDED TO DECOLOURIZE DCPIP / DROPS
Brand A (long-life)	20
Brand B	13
Brand C	6
Freshly squeezed	5
0.1% ascorbic acid	10

a. Use the following formula to calculate the vitamin C in Brand A.

% vitamin C in Brand A =

$$\frac{(\text{number of drops of 0.1\% ascorbic acid}) \times 0.1}{\text{number of drops of Brand A}} = \ldots\ldots\%$$

(2)

b. Which brand of shop-bought orange juice had the highest vitamin C content? (1)
c. Long-life orange juice is heated to a high temperature, and then sealed in cartons. What effect does this have on the vitamin C content? (2)
d. Give **one** important function of vitamin C in humans. (1)
e. Name **one** other source of vitamin C apart from fruit juice. (1)

3. The table shows the amounts of four nutrients required by four people for a balanced diet.

PERSON	PROTEIN / g	IRON / mg	CALCIUM / mg	VITAMIN C / mg
14 year-old boy	66	11	700	25
14 year-old girl	55	13	700	25
30 year-old woman	53	12	500	30
30 year-old pregnant woman	60	14	1200	60

a. i. Explain why there is a difference in the amount of protein required by the 14 year-old boy and the 30 year-old woman. (3)
   ii. Explain why there is a difference in the amount of iron required by the 14 year-old girl and the 14 year-old boy. (2)
   iii. Explain why there is a difference in the amount of calcium required by the two 30 year-old women. (2)
b. State the role of vitamin C in the human body. (1)

Cambridge IGCSE Biology 0610 Paper 2 Q6 June 2006

Supplement

4. A bioreactor may be set up in such a way that a fungus, *Fusarium graminearum*, can be grown on waste materials from the flour and paper industries. The fungus is harvested and processed to make fibres that are dried and pressed to produce **mycoprotein**. Mycoprotein may be flavoured and sold under the trade name Quorn. The table below provides some nutritional information about a variety of foods.

CONTENT PER 100 g	QUORN	STEAK	CHICKEN	OLIVE OIL	POTATO
Protein / g	12.0	31	24.8	0	4.1
Fat / g	3.2	10.5	5.4	220	0.3
Carbohydrate / g	1.0	0.5	0.2	0	32.4
Cholesterol / g	0	80	74	0	0
Dietary fibre / g	5.1	0	0	0	2.8
Energy / kJ	355	950	620	3600	575

a. Use information from the table to explain why Quorn is regarded as a healthy and nutritious food. (4)

b. Jack produced a meal by frying 250 g of Quorn in 10 g of olive oil, and adding a baked potato. Calculate the total energy value of the meal. Show all of your working. (4)

5. Simple chemical tests on foods can show which type of nutrients they contain.

A group of students was given samples of four different foods to test – the samples were prepared by dissolving powdered food in distilled water. The four foods were identified as A, B, C and D. They were also given a sample of distilled water. They carried out three tests on the five samples. Test 1 was for glucose, test 2 was for protein and test 3 was for starch. The results of the tests are shown in the table.

FOOD TEST	1: GLUCOSE	2: PROTEIN	3: STARCH
Sample A	Blue	Purple	Brown
Sample B	Orange	Blue	Brown
Sample C	Blue	Blue	Black
Sample D	Orange	Purple	Brown
Distilled water	Blue	Blue	Brown

a. Which colour indicates the presence of starch? (1)
b. What is the name of the solution that gives a purple colour if protein is present? (1)
c. Which one of the samples contained protein and glucose but no starch? (1)
d. Why were the tests also carried out on distilled water? (1)
e. This series of tests does not detect the presence of fat. Describe a test for fat, including a positive result for the test. (3)
f. Explain **one** danger to health of eating too many foods that contain large amounts of fat. (1)

THE HUMAN DIET: Crossword

ACROSS:
1 Deficiency disease that results from a lack of vitamin C in the diet
3 Tasty – but too much can raise blood pressure
4 Mineral required for healthy teeth and bones
7 Should form the bulk of a balanced diet
9 A type of lipid used to produce some hormones
10 The body-building food
11 Essential for insulation and as an energy store
13 Process that releases energy from foods
14 This type of diet has all the essential foods in the correct proportions
15 Mineral component of haemoglobin – deficiency causes anaemia
16 11 across is needed to make this cell boundary

Down
2 Deficiency disease due to low levels of vitamin D
3 Sweet carbohydrate for tea and coffee drinkers
5 Lipid that can cause damage to blood vessels
6 High in protein and dietary fibre – an ideal meal when eaten on toast! (5,5)
8 The result of a poorly balanced diet
12 Essential for the correct formation of faeces

Human digestive system

Digestion is the breaking down of large insoluble food molecules into small soluble molecules using mechanical and chemical processes so that they can be absorbed into the bloodstream.

THE DIGESTIVE SYSTEM

Food is digested in the **alimentary canal**. This is a long tube that starts at the mouth, runs through the stomach and intestines and finishes at the anus.

Food is broken down with the help of digestive juices, which contain special chemicals called **enzymes**.

DEALING WITH FOOD

There are five stages in the way we deal with food:

1 **Ingestion** – taking food into the body through the mouth ('eating')

2 **Digestion** – the breakdown of large, insoluble food molecules into small, water-soluble molecules using mechanical and chemical processes

3 **Absorption** – moving digested molecules from the alimentary canal into the bloodstream or the lymph so they can be transported around the body

4 **Assimilation** – movement of absorbed food molecules into cells where they are used and become part of the cells

5 **Egestion** – getting rid of food that could not be digested (e.g. dietary fibre) by passing it as faeces

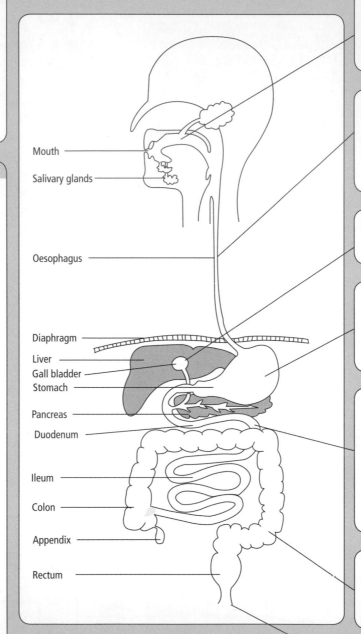

Mouth

Salivary glands

Oesophagus

Diaphragm

Liver

Gall bladder

Stomach

Pancreas

Duodenum

Ileum

Colon

Appendix

Rectum

Supplement

Diarrhoea is the loss of watery faeces. It can occur during CHOLERA, a disease caused by a bacterium. This disease causes osmosis into the gut leading to dehydration of the body and loss of salts from the blood.

Mouth – Mechanical digestion takes place here. The teeth cut and grind the food, which is mixed with saliva. The enzyme salivary amylase breaks starch down into maltose (sugar).

Oesophagus – lumps of moist, chewed up food are carried to the stomach by muscle movements. This is called **peristalsis** and also moves partly-digested food along the small intestine.

Gall bladder – stores bile used to help in the digestion of fats.

Stomach – the stomach is like a sack. Here the enzyme **pepsin** breaks big proteins down into small proteins (**polypeptides**). This can take several hours.

Small intestine – this is made up of the **duodenum** and the **ileum**. Here the enzymes pancreatic amylase, trypsin and lipase break down starches, fats, proteins and complex sugars into small soluble molecules.

Fully digested food is absorbed into the bloodstream.

Large intestine – this is made up of the **colon** and the **rectum**. Only undigested food reaches here. Water is absorbed.

Anus – undigested solid food is passed out as **faeces**.

Digestion, absorption and assimilation

Large food molecules are broken down in our digestive system mostly by chemical reactions. The products are **absorbed** into the blood or lymph then **assimilated** into cells.

Supplement

Supplement

Supplement

ENZYMES and DIGESTION

ENZYME	ACTION	SPECIAL CONDITIONS
Amylase in saliva and pancreatic juice	Starch ⟶ maltose	Alkaline (hydrogen carbonate in saliva)
Maltase on lining of small intestine	Maltose ⟶ glucose	Alkaline (hydrogen carbonate in pancreatic juice)
Pepsin (a protease)	Protein ⟶ peptides	Acidic – pH 2 (hydrochloric acid, which also denatures enzymes in harmful microbes)
Trypsin (a protease)	Peptides ⟶ amino acids	Alkaline (as for maltase)
Lipase	Fats ⟶ fatty acids + glycerol	Alkaline (as for maltase)

BILE and EMULSIFICATION
- made in the liver and stored on the gall bladder
- emulsifies fats – converts large globules into smaller droplets

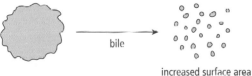

bile

increased surface area for the action of lipase

- contains hydrogen carbonate which neutralises the acidic chyme from the stomach

ABSORPTION IN THE SMALL INTESTINE

The wall of the small intestine has millions of tiny finger-shaped structures called **villi**. These give a huge surface area for absorbing digested food easily.

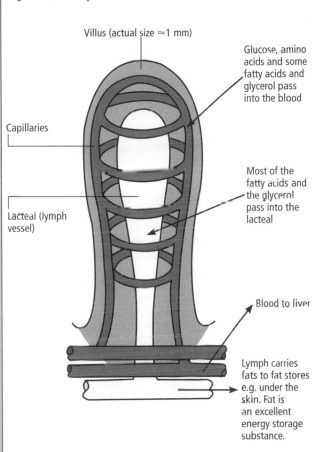

Villus (actual size ≈1 mm)

Glucose, amino acids and some fatty acids and glycerol pass into the blood

Capillaries

Most of the fatty acids and the glycerol pass into the lacteal

Lacteal (lymph vessel)

Blood to liver

Lymph carries fats to fat stores e.g. under the skin. Fat is an excellent energy storage substance.

DIGESTED FOOD AND THE LIVER

The liver is the largest organ in the human body. It is like a 'chemical factory'. It processes digested food and other substances in the blood. Its main functions are:
- storing vitamins, minerals and glycogen
- producing special chemicals, e.g. bile for digestion
- processing unwanted substances from the blood, e.g. alcohol is removed from the blood by the liver and made into a less toxic substance
- generating heat to keep the body's internal temperature at 37 °C
- excess amino acids are **deaminated** to give urea for excretion from the kidneys (see p. 73).

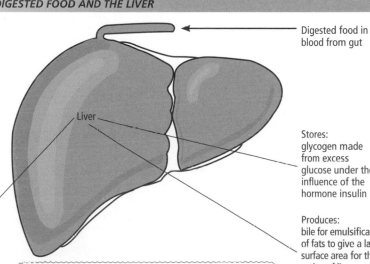

Digested food in blood from gut

Liver

Stores: glycogen made from excess glucose under the influence of the hormone insulin

Produces: bile for emulsification of fats to give a larger surface area for the action of lipase fibrinogen for clotting of blood

The liver is severely damaged by excess alcohol. In the disease **cirrhosis** the liver cells cannot perform their normal functions

TEETH AND TOOTH DECAY

The teeth in the skull

Upper jaw

Incisors – for cutting and biting

Canine – for holding and cutting

Lower jaw

Premolars – for chewing and crushing

Molars (third one is hidden) – for chewing and crushing

CHEWING (MASTICATION) AND SWALLOWING
- the first stage in preparing food for the alimentary canal
- food is cut up by the *teeth* and mixed with *saliva* by the tongue
- the ball of food (the *bolus*) is pushed to the back of the mouth
- swallowing - forcing the bolus into the oesophagus - is a *voluntary action*
- at the same time a *reflex action* lifts the *epiglottis* (a flap of tissue) across the entrance to the larynx, so food does not enter the breathing passages.

A MOLAR TOOTH

Enamel – the hardest tissue in the body. Produced by **tooth-forming cells** and made of calcium salts. Once formed, enamel cannot be renewed or extended.

Cement – similar in composition to dentine, but without any canals. It helps anchor the tooth to the jaw.

Crown

Root embedded in jawbone

Pulp cavity contains:
- **tooth-producing cells**
- **blood vessels**
- **nerve endings** which can detect pain.

Dentine – forms the major part of the tooth. Harder than bone and made of calcium salts deposited on a framework of **collagen fibres**. The dentine contains a series of fine canals which extend to the pulp cavity.

Gum – usually covers the junction between enamel and cement. The gums recede with age.

CARE OF THE TEETH
- reduce number of sweet or acidic foods - fizzy drinks are especially harmful!
- brush to remove sticky food remains that may begin the build-up of *plaque*
- use dental floss to remove material from between teeth

PLAQUE CAUSES TOOTH DECAY

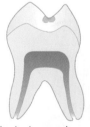

Decay begins in enamel – **no pain**.

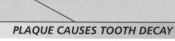
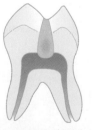

Decay penetrates dentine and reaches pulp – **severe toothache**.

Bacteria now infect pulp and may form abscess at base of tooth – **excruciating pain**.

Fluoride in toothpaste or added to water may stop the decay of teeth, but
- if added to drinking water we can't control how much we take in
- some teeth develop brown spots when contacted by fluoride - this can be unsightly and reduce confidence.

1. The diagram below shows the human alimentary canal.

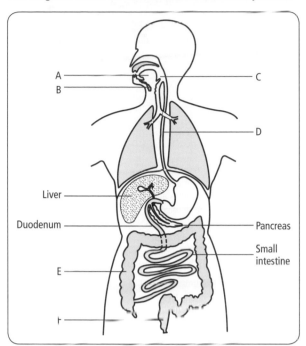

a. Name the part labelled **A**. (1)
b. Explain how **A** and **B** work together to make swallowing more efficient. (2)
c. Why is **C** important when swallowing? (1)
d. Draw a labelled diagram to explain how food is moved along the structure labelled **D**. (2)
e. Describe **two** features of the small intestine that help to increase its surface area. (2)
f. The liver has many functions, including the storage of excess carbohydrate.
   i. Name the carbohydrate used to store excess sugar in the liver. (1)
   ii. Name the blood vessel that carries sugar from the intestine to the liver. (1)
   iii. Name the blood vessel that carries glucose released from the liver some time after feeding. (1)
   iv. The liver also produces bile. What is the function of bile? Where does bile have its effect? (2)
g. The pancreas produces enzymes and releases them into the duodenum.
   i. How do the enzymes produced in the pancreas get into the duodenum? (1)
   ii. Name one enzyme produced by the pancreas. State what kind of foodstuff it works on, and what the products are. (3)
h. Give **one** function of each of the parts labelled **E** and **F**. (2)

2. Lengths of Visking tubing were set up as shown in the diagram.

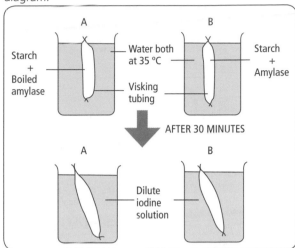

Tube A contained starch and boiled amylase solution
Tube B contained starch and amylase.
Both tubes were left in a water bath at 35 °C for 30 minutes. The tubes were then removed from the water baths, gently blotted with paper towels, and then placed in beakers of dilute iodine solution for five minutes. Visking tubing is permeable to iodine solution.

a. What colour would you expect to see inside A and inside B? (2)
b. Explain why the results are different for tubings A and B. (2)
c. Describe a test you could use to prove your explanation. (3)
d. The drawing below shows a section through some villi.

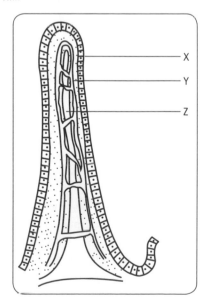

   i. Identify the parts labelled **X**, **Y** and **Z**. (3)
   ii. Where in the alimentary canal would villi be found? (1)

iii. Which of the parts **X**, **Y** or **Z** is represented by the Visking tubing? (1)

iv. What is the function of the part labelled **Z**? (1)

3. The diagram below shows the teeth in the lower jaw of an adult human.

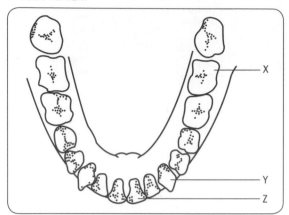

a. i. Name the teeth labelled **X**, **Y** and **Z**. (3)

ii. Describe the functions of teeth **X** and **Z**. (2)

b. Name **one** mineral and **one** vitamin that are essential for the healthy development of teeth. (2)

c. The diagram below shows a section through a tooth.

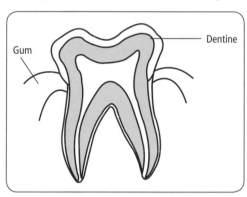

i. Tooth decay is caused by bacteria getting into the dentine. Explain how bacteria can enter the dentine. (3)

ii. List three actions you could take to reduce the risk of tooth decay. (3)

Cambridge IGCSE Biology 0610 Paper 2 Q6 November 2006

4. The diagram shows the apparatus used to investigate the digestion of milk fat by an enzyme. The reaction mixture contained milk, sodium carbonate solution (an alkali) and the enzyme. In Experiment 1, bile was also added. In Experiment 2, an equal volume of water replaced the bile. In each experiment, the pH was recorded at 2-minute intervals.

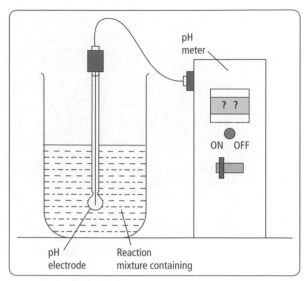

	Either: Experiment 1	**or: Experiment 2**
	milk (contains fat)	milk (contains fat)
	sodium carbonate solution	sodium carbonate solution
	bile	water
	enzyme	enzyme

The results of the two experiments are given in the table.

TIME / MINUTES	pH	
	EXPERIMENT 1: WITH BILE	EXPERIMENT 2: NO BILE
0	9.0	9.0
2	8.8	9.0
4	8.7	9.0
6	8.1	8.8
8	7.7	8.6
10	7.6	8.2

a. Milk fat is a type of lipid. Give the name of an enzyme which catalyses the breakdown of lipids. (1)

b. What was produced in each experiment to cause the fall in pH? (1)

c. i. For Experiment 1, calculate the average rate of fall in pH per minute, between 4 minutes and 8 minutes. Show clearly how you work out your final answer. (2)

ii. Why was the fall in pH faster when bile was present? (1)

5. a. i. What name is given to the type of enzyme that breaks down protein in the diet? (1)
ii. What is the product formed when protein is broken down in this way? (1)
b. This table shows the effect of pH on the activity of this type of enzyme.

pH	1.0	2.0	3.0	4.0	5.0	6.0
RATE OF FORMATION OF PRODUCT / MMOLES PER MINUTE	9.5	22.8	16.4	8.1	2.5	0.0

Plot these results in a suitable graph. Supply a title for the graph. (5)
c. Suggest in which part of the human digestive system this enzyme would be working. (1)
d. Why is it necessary to break down large molecules such as protein in the digestive system? (2)

REVISION SUMMARY: Match the term to the correct definition

	TERM		DESCRIPTION
A	Absorption	1	The conversion of fat globules into smaller droplets
B	Amylase	2	The region of the intestine where most digested food is absorbed
C	Anal sphincter	3	Link between the gall bladder and the small intestine
D	Assimilation	4	Faeces are stored here before they are expelled
E	Bile duct	5	Controls the amount of food leaving the stomach
F	Colon	6	Secretes several enzymes into the small intestine
G	Digestion	7	Muscular contractions that move a bolus of food along the gut
H	Emulsification	8	A protein-digesting enzyme
I	Faeces	9	The organ that produces bile
J	Gall bladder	10	Structures that increase the surface area of the ileum
K	Ileum	11	Region of the intestine in which protein digestion begins
L	Lipase	12	The transfer of digested food from the gut to the bloodstream
M	Liver	13	Region from which most water is absorbed from the contents of the intestine
N	Mastication	14	A fat-digesting enzyme
O	Pancreas	15	A mixture of water, enzymes and mucus
P	Pepsin	16	The chewing and mixing of food with saliva
Q	Peristalsis	17	The use of food molecules in the body
R	Pyloric sphincter	18	The site for bile storage
S	Rectum	19	The breakdown of food into small, soluble molecules ready for absorption
T	Saliva	20	Undigested and waste products expelled from the body
U	Stomach	21	Controls the passing of faeces from the body
V	Villi	22	An enzyme that digests starch to maltose

Chapter 13:
Photosynthesis and plant nutrition

Plants make food by photosynthesis

TISSUES IN THE LEAF OF A DICOTYLEDONOUS PLANT

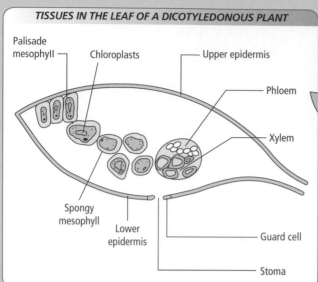

Palisade mesophyll
Chloroplasts
Upper epidermis
Phloem
Xylem
Spongy mesophyll
Lower epidermis
Guard cell
Stoma

The occurrence of photosynthesis can be demonstrated by testing for the presence of starch (made from glucose).

$$STARCH \xrightarrow[\text{SOLUTION}]{\text{IODINE}} \text{INTENSE BLUE-BLACK COLOURATION}$$

The rate of photosynthesis can be measured by collecting evolved oxygen, usually from the cut end of the aquatic plant, *Elodea*.

EQUATION FOR PHOTOSYNTHESIS

$$CARBON\ DIOXIDE + WATER \xrightarrow[\substack{CHLOROPHYLL \\ ENZYMES}]{LIGHT\ ENERGY} GLUCOSE + OXYGEN$$

$$6CO_2 + 6H_2O \xrightarrow[\text{CHLOROPHYLL}]{\text{LIGHT}} C_6H_{12}O_6 + 6O_2$$

PLANTS NEED MINERALS TOO!

MINERAL	SOURCE	IMPORTANCE
Magnesium	Absorbed as Mg^{2+} from the soil solution.	Manufacture of chlorophyll: absence makes leaves turn yellow and eventually stops photosynthesis.
Nitrogen	Absorbed as nitrate (NO_3^-) or ammonium (NH_4^+) from the soil solution.	Manufacture of proteins, nucleic acids and plant hormones: absence causes poor growth, especially of leaves. This is an important limiting factor, so many farmers add nitrogen-containing fertilisers to their land.

Hydrocarbonate indicator can be used to study gas exchange in an aquatic plant.

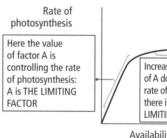

APHIDS TAP PHLOEM FOR THE SUGARS MADE BY PHOTOSYNTHESIS

YUMMY!

LIMITING FACTORS AND PHOTOSYNTHESIS

There are a number of conditions that must be satisfied for photosynthesis to proceed. The factor that **is nearest to its minimum value** will limit the rate of this process.

Rate of photosynthesis

Here the value of factor A is controlling the rate of photosynthesis: A is THE LIMITING FACTOR

Increase in availability of A does not increase rate of photosynthesis: there is some other LIMITING FACTOR

Availability of factor A

Limiting factors in photosynthesis are **carbon dioxide concentration, light intensity and temperature** (affects the enzymes that catalyse the chemical reactions of photosynthesis).

WHAT HAPPENS TO THE GLUCOSE?

Glucose

- Respiration to provide the energy to drive the metabolic reactions needed to keep the plant alive.
- Conversion to other molecules such as oils and proteins. This may require mineral salts (e.g. nitrate).
- Conversion to sucrose for transport to other parts of the plant via the phloem (translocation).
- Conversion to cellulose for the construction of plant cell walls.
- Conversion to starch for storage. Starch is insoluble so does not affect water potential of plant cells.

1. The diagram shows the apparatus used to investigate the effect of light intensity on the rate of photosynthesis.

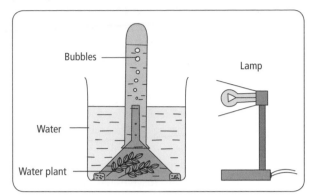

a. How would the light intensity be varied? (1)
b. Suggest a suitable control for the investigation. (1)
c. Name **two** factors that should be kept constant if this is to be a fair test. (2)
d. The rate of photosynthesis was measured by counting the number of bubbles released into the test tube every 5 minutes. The results obtained are shown below.

LIGHT INTENSITY / ARBITRARY UNITS	RATE OF PHOTOSYNTHESIS / NUMBER OF BUBBLES PER 5 MINUTE PERIOD
0	1
10	8
20	15
30	23
40	30
50	37
60	43
70	44
80	44

i. Plot a graph of these results. (5)
ii. Use your graph to find the number of gas bubbles given off in a five minute period at a light intensity of 35 units. (2)
iii. Describe the relationship between light intensity and rate of photosynthesis. Explain this relationship. (4)
iv. Suggest **two** ways in which the bubbles of gas differ from normal atmospheric air. (2)
v. Suggest **one** way in which the results of the experiment could be made more reliable. (1)
e. Two students remembered that the school aquarium had red-coloured lighting. They were interested in whether the colour (wavelength) of light would affect the rate of photosynthesis. Describe an experiment that they could carry out to test the hypothesis that colour of light affects the rate of photosynthesis. Name the input variable, the outcome variable and any fixed variables in their experiment. (5)

2. The graph shown below indicates the amount of sugar contained in the leaves of a group of plants, kept in a greenhouse, over a seven-day period.

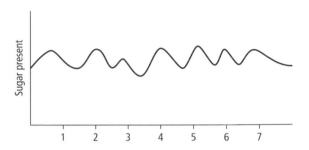

a. Write a word equation for the process that produces the sugar in the plants. (4)
b. Name the process shown by the equation. (1)
c. Why does the sugar level rise and fall during the day? (2)
d. Suggest a reason for the lower peak on day 3. (2)
e. The greenhouse owner used a paraffin burner during the spring. Give **two** reasons why using the burner might increase the yield of sugar produced. (2)

3. The diagram shows a section through part of a leaf.

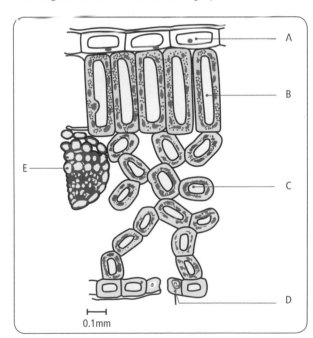

a. i. Name the cells labelled **A, B, C, D** and **E**. (5)
ii. In which of these cells does most photosynthesis occur? (1)
iii. Which of these cells would transport water and minerals to the leaf? (1)

b. In what form is carbohydrate transported around the plant? (1)

c. i. In what form is carbohydrate stored in an onion? (1)

ii. Describe how you would carry out a test for this storage carbohydrate. Describe a positive result. (2)

iii. Describe two uses, other than storage, for the carbohydrate made by the process of photosynthesis. (2)

d. Look back at the leaf section. Use the scale to work out

i. the thickness of the leaf

ii. the length of a palisade cell

Show your working. (2)

4. Three plants were grown to study the effects of nitrate and magnesium ion deficiency on their development. They were kept in the same conditions, except for the types of minerals supplied.

Plant **A** was provided with all essential minerals.

Plant **B** was given all minerals except nitrate ions.

Plant **C** was given all minerals except magnesium ions.

The diagram shows the plants a few weeks later.

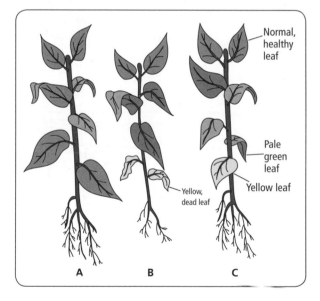

a. State **three** conditions, **other than** water and the concentration of mineral ions, that would need to be kept the same for all the plants, in order to make the investigation a fair test. (3)

b. Describe and explain the effect on plant growth of

i. a deficiency of nitrate ions on plant **B**;
description
explanation (4)

ii. a deficiency of magnesium ions on plant **C**.
description
explanation (2)

c. A farmer tested the soil in a field and found that there was a high nitrate ion concentration.
The farmer then grew a crop in this field.
After the crop was removed, the soil was tested again. The nitrate ion concentration had decreased.

i. Suggest **two** reasons why the nitrate ion concentration had decreased. (2)

ii. Describe **two** methods the farmer could use to improve the nitrate ion concentration in the soil. (2)

d. Some species of plant grow well in soil that is always low in nitrate ions.
Explain how they can obtain a source of nitrogen compounds. (3)

Cambridge IGCSE Biology 0610 Paper 3 Q1 June 2005

Supplement

REVISION SUMMARY: Match the term to the correct definiton

	TERM		DEFINITION
A	Limiting factor	1	The molecule that absorbs light energy to drive photosynthesis
B	Nitrate	2	The form in which carbohydrate is transported away from the leaves
C	Magnesium	3	Are needed to catalyse the chemical reactions in photosynthesis
D	Chlorophyll	4	Some aspect of the environment needed for photosynthesis that is nearest to its minimum value
E	Palisade cell	5	Gas that is a raw material for photosynthesis
F	Stomata	6	The plant tissue that transports the products of photosynthesis away from the leaves
G	Carbon dioxide	7	The organelle in which photosynthesis takes place
H	Oxygen	8	Pores that allow the entry of carbon dioxide
I	Light	9	A molecule made from glucose that is essential for plant cell structure
J	Phloem	10	Mineral ion required for protein synthesis
K	Sucrose	11	Storage carbohydrate in plants
L	Cellulose	12	The source of energy for photosynthesis
M	Starch	13	Adapted for its function by being packed with chloroplasts
N	Chloroplast	14	Gaseous product of photosynthesis essential for aerobic respiration
O	Enzymes	15	Mineral ion essential for the production of chlorophyll

Chapter 14:
Plant transport

Water movement through a plant
begins with evaporation from the leaf surface.

WATER UPTAKE BY PLANTS

can be measured using a bubble potometer.

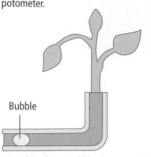

Bubble

Water loss draws bubble along capillary tube as water uptake replaces loss.

Transpiration is evaporation of water from mesophyll cells followed by loss of water through the stomata. This occurs by evaporation and lowers the water potential in the tissues of the leaf.

Stomata are essential if the uptake of CO_2 for photosynthesis is to go on, but if stomata are open water evaporated from the spongy mesophyll can diffuse out of the leaf down the gradient of water potential.

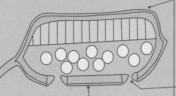

LEAF STRUCTURE MAY REDUCE TRANSPIRATION.

* thick, waxy cuticle reduces evaporation from epidermis.
* stomata may be sunk into pits, which trap a pocket of humid air.
* leaves may be rolled with the stomata on the inner surface close to a trapped layer of humid air.

HUMID AIR

* leaves may be needle-shaped to reduce their surface area.
* plants adapted to save water in dry environments are called **xerophytes**.

Water moves from xylem to enter leaf tissues down water potential gradient.

Water moves up the stem in the xylem due to the **tension** caused by water loss from the leaves.

98% of the water taken up by a plant is 'lost' to the atmosphere by transpiration!

WATER IS IMPORTANT
* it is a substrate for photosynthesis
* it is a pathway for transport of mineral ions through the plant
* it provides support, through turgidity
* evaporation may help to cool leaf surfaces

Water uptake occurs by osmosis from the soil solution (high water potential) into the root cells (lower water potential).

ATMOSPHERIC CONDITIONS MAY AFFECT TRANSPIRATION
* **Wind** moves humid air away from the leaf surface and increases transpiration.
* **High temperatures** increase the water holding capacity of the air and increase transpiration.
* **Low humidity** increases the water potential gradient between leaf and atmosphere and increases transpiration.
* **High light intensity** causes stomata to open (to allow photosynthesis), which allows transpiration to occur.

WATER ENTERS THE PLANT THROUGH THE ROOTS

flow of water and minerals to the stem

soil particles

root hair (large surface area helps uptake of water and minerals)

root xylem vessels

water and minerals absorbed by root hair

TRANSPIRATION IS CONTROLLED BY STOMATA

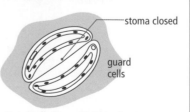

stoma open

guard cells

stoma closed

guard cells

Guard cells swollen with water.
Stoma (pore) is held open.
Water can be lost through the stoma.

Guard cells limp with lack of water.
Stoma (pore) is closed.
Little water is lost through the stoma.

Transport systems in plants

move vital substances from **sources** (sites of uptake or manufacture) to **sinks** (sites of production or storage).

INVESTIGATION OF FUNCTIONS OF PLANT TISSUES
- **Phloem** aphids (greenfly) can sample contents and radioactive sugars identify tissue as site of sugar transport.
- **Xylem** water-soluble dyes trace pathway of water movement
- **Roots** inhibitors of respiration stop active uptake of ions

FRUITS AND GROWING POINTS these are **sinks** for many nutrients.

Fruits
- demand water for swelling of ovary wall if succulent
- demand sucrose to be converted to starch as an energy store

Growing points
- demand water for cell swelling
- demand sucrose as energy source for cell division
- demand all nutrients as raw materials for cell production.

STEM the position of the vascular bundles (in a ring with soft cortex in the centre) helps to support the stem against sideways forces, e.g. wind.

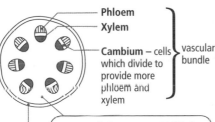

Phloem
Xylem
Cambium – cells which divide to provide more phloem and xylem } vascular bundle

Cortex cells become turgid (filled with water) and help to support non-woody parts.

Epidermis protective against, for example, infection by viruses and bacteria, and dehydration.

PHLOEM transport of organic products of photosynthesis i.e. sugars (carried as sucrose) and amino acids. This is TRANSLOCATION. Aphids (greenfly) can sample contents and show pathway of sucrose movement.

LEAVES are both **sinks** and **sources**.

Sinks
- water as a reactant in photosynthesis
- magnesium as a component of the chlorophyll molecule

Sources
- glucose formed during photosynthesis
CARBON DIOXIDE + WATER
↓
GLUCOSE + OXYGEN

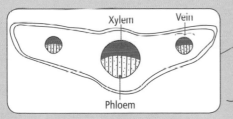

Xylem   Vein

Phloem

NOTICE Sugar can move up and down phloem at the same time

XYLEM transport of water and dissolved mineral salts - movement is always **up** the stem. This is part of TRANSPIRATION. Water soluble dyes can be used to trace the pathway of water movement.

DIRECTION OF TRANSPORT VARIES WITH THE SEASONS! Sucrose will be transported **from** stores in the root **to** leaves in spring, but **to** stores in the root **from** photosynthesizing leaves in the summer and early autumn.

ROOT

Root hair extended cells of epidermis increase surface area for water/ion uptake.

Epidermis protective, e.g. against infection by fungi.

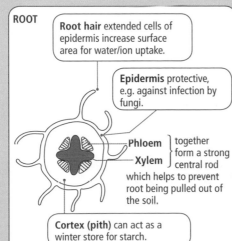

Phloem
Xylem } together form a strong central rod which helps to prevent root being pulled out of the soil.

Cortex (pith) can act as a winter store for starch.

ROOTS: are both **sinks** and **sources**.

Sinks
- sucrose to supply energy for growth and active uptake of ions from the soil
- sucrose to be converted to starch for storage

Sources
- water, absorbed from the soil solution
- ions, absorbed from soil by active transport
- sucrose, before leaves are capable of photosynthesis, in spring.

1. Using the apparatus shown below, a student set out to investigate the factors that influence water uptake by a plant.

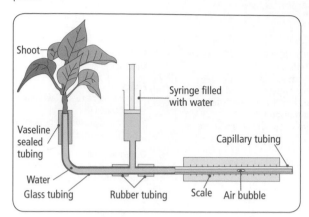

After a series of experiments he obtained the following results.

ENVIRONMENTAL CONDITION	TIME TAKEN FOR BUBBLE TO MOVE 10 CM / MIN	RATE OF BUBBLE MOVEMENT / CM PER MINUTE
1 High light intensity	5	
2 High humidity (plant enclosed in clear plastic bag)	16	
3 Wind (electric fan blowing over plant surface)	1	
4 Dark and windy	17	
5 Dark and low humidity	25	

a. Calculate the **rate** at which the bubble moves. Plot these results in the form of a bar chart. (5,5)

b. Plants require light for photosynthesis, and 'anticipate' the need for carbon dioxide uptake during photosynthesis by opening their stomata under appropriate conditions. Does this help to explain the results of the first experiment? Explain your answer. (2)

c. Water is lost from leaves by evaporation and diffusion **if the stomata are open, and a suitable water potential gradient exists**. Use this information to explain the results of experiments 2 and 5. (3)

d. How can you explain the results of experiments 3 and 4? (2)

2. Four leaves were removed from the same plant. Petroleum jelly (a waterproofing agent) was spread onto some of the leaves, as follows:

Leaf **A**: on both surfaces
Leaf **B**: on the lower surface only
Leaf **C**: on the upper surface only
Leaf **D**: none applied

Each leaf was then placed in a separate beaker, as shown in the diagram.

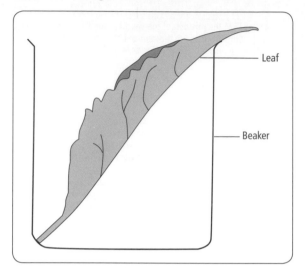

Each beaker was weighed at intervals. The results are shown in the graph.

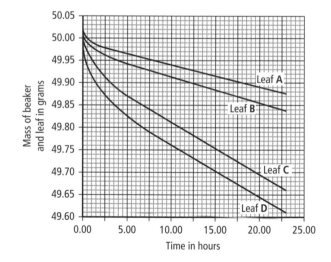

a. Give evidence from the graph in answering the following questions.
   i. Which surface (upper or lower) loses water most rapidly? (1)
   ii. Is water lost from both surfaces of the leaf? (1)

b. The diagrams below show the appearance of each surface of the leaf as seen through a microscope.

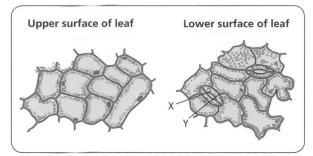

Upper surface of leaf	Lower surface of leaf

i. Name space **X** and cell **Y**. (2)
ii. Use information in the diagram to explain why the results are different for leaves **B** and **C**. (2)

3. The diagrams show the underside of two leaves.

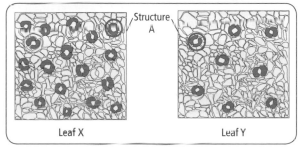

Leaf X   Leaf Y

a. Name the structures labelled **A**. (1)
b. Make a labelled drawing of a side view of one of these structures. (3)
c. Water can be lost from a leaf through these structures. The water on the surface of spongy mesophyll cells changes to water vapour, and then moves out of the leaf down a concentration gradient.
　i. Name the process in which water changes from liquid to vapour. (1)
　ii. Name the process by which the molecules of water vapour move down a concentration gradient. (1)
　iii. Suggest one advantage and one disadvantage of this process to the plant. (2)
d. The view of the leaf underside covers 1 mm². Each of the leaves measured 5 cm × 2 cm. Calculate the total number of the structures **A** on each of the leaves. Show your working. (4)

LEAF	NUMBER OF STRUCTURES A IN 1 MM²	TOTAL NUMBERS OF STRUCTURES A ON LEAF
X		
Y		

e. Which of the leaves is better adapted to living in dry, hot conditions? Explain your answer. (2)
f. Give two other features of leaves adapted to life in dry, hot conditions. (2)

4. The diagram on the left below shows a transverse section through an *Ammophila* leaf. This plant has very long roots. The diagram on the right shows a cactus plant. Both plants live in very dry conditions.

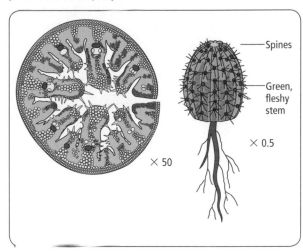

a. Suggest how each of the following adaptations would enable the named plant to survive in very dry conditions.
　i. *Ammophila*
　　1. rolled leaves with stomata on the inside of the leaf (2)
　　2. thick waxy cuticle on the outside of the leaf (1)
　ii. Cactus
　　1. very long roots (1)
　　2. fleshy green stem (2)
b. Suggest why having only a few, very small leaves could be a disadvantage to a plant. (2)
c. Water is involved in a number of processes in plants.
　Complete the table by
　i. naming the processes described;
　ii. stating one variable that, if increased, would speed up the process.

DESCRIPTION OF PROCESS	NAME OF PROCESS	VARIABLE THAT, IF INCREASED, WOULD SPEED UP THE PROCESS
Absorption of water from the soil		
Using water to form glucose		
Movement of water vapour out of leaves		

(6)

Cambridge IGCSE Biology 0610 Paper 3 Q4 June 2005

REVISION SUMMARY: Fill in the missing words

Use words from the following list to complete the paragraphs below.

PHLOEM, VASCULAR, ACTIVE TRANSPORT, OSMOSIS, DIFFUSION, RESPIRATION, XYLEM, SURFACE AREA, NITRATE, IONS, SOLVENT, PHOTOSYNTHESIS, DIGESTION, SUPPORT, HAIRS, EPIDERMIS

Water is obtained by plants from the soil solution. The water enters by the process of ...., via special structures on the outside of the root called root .... These structures increase the .... of the root and, as well as absorbing water, they can also take up .... such as ...., which is required for the synthesis of chlorophyll. These substances are absorbed both by .... and by ...., a process that requires the supply of energy.

Plant cells rely on water for ...., as a .... and as a raw material for .... Water is used as a transport medium for both ions and sugars. The ions are transported, along with water, in the .... tissue. Sugars are transported through living cells of the .... and these specialized tissues are grouped together into .... bundles. (13)

Chapter 15:
Transport in animals

Blood and the human circulatory system

The circulatory system is a system of blood vessels with a pump and valves to ensure one-way flow of blood.

Blood tissue transports vital substances around the body. It also plays an essential part in protecting the body from damage and disease.

THE HEART

Blood is pumped around the body by a muscular pump called the **heart**. Blood leaves the heart via **arteries** and enters via **veins**.

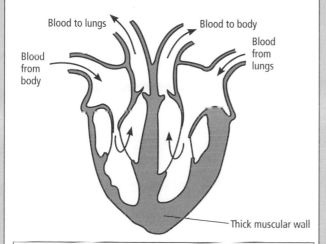

Blood to lungs

Blood to body

Blood from lungs

Blood from body

Thick muscular wall

- Blood passes through the heart twice for each trip around the body. We call this the **double circulation**. This means blood is delivered at **high pressure** to all of the organs.
- **Valves** in the heart prevent blood from being pushed backwards up into the atria when the heart 'beats'.
- Heart activity can be checked by ECG, pulse rate and listening to sounds of valves closing.

BLOOD TISSUE

Spinning blood in a centrifuge separates the plasma from the cells.

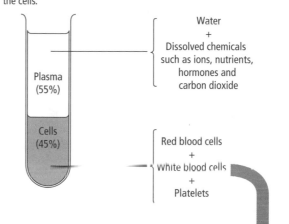

Plasma (55%)

Cells (45%)

Water + Dissolved chemicals such as ions, nutrients, hormones and carbon dioxide

Red blood cells + White blood cells + Platelets

THE FUNCTIONS OF BLOOD CELLS

Red blood cells carry oxygen by combining it with **haemoglobin**.

White blood cells fight disease by producing **antibodies** or by phagocytosis (**engulfing** dangerous microbes).

Platelets are cell fragments. They help **clotting**, preventing blood loss and infection.

THE HUMAN CIRCULATORY SYSTEM

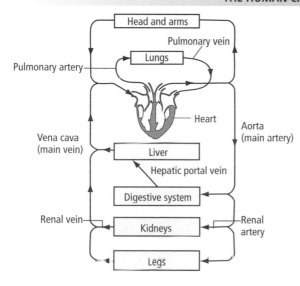

Head and arms

Pulmonary vein

Pulmonary artery

Lungs

Heart

Aorta (main artery)

Vena cava (main vein)

Liver

Hepatic portal vein

Digestive system

Renal vein

Kidneys

Renal artery

Legs

- **Arteries** carry blood **away** from the heart. Blood in arteries has more oxygen than blood in veins – except for the pulmonary artery and veins, which go to and from the lungs!
- **Arteries** have thick walls to withstand higher blood pressure.
- **Veins** have thinner walls but have valves to stop blood flowing 'backwards'.
- **Veins** carry blood back to the heart.
- **Capillaries** reach all tissues and organs.

CORONARY HEART DISEASE affects 1 in 4 people between the ages of 30 and 60. High risk factors include
- Cigarette smoking
- High blood cholesterol
- Obesity
- Lack of regular exercise

A 'heart attack' involves death of an area of heart tissue following an interrupted blood supply – usually when the coronary artery (which delivers blood to heart muscle itself) is narrowed by fatty deposits.

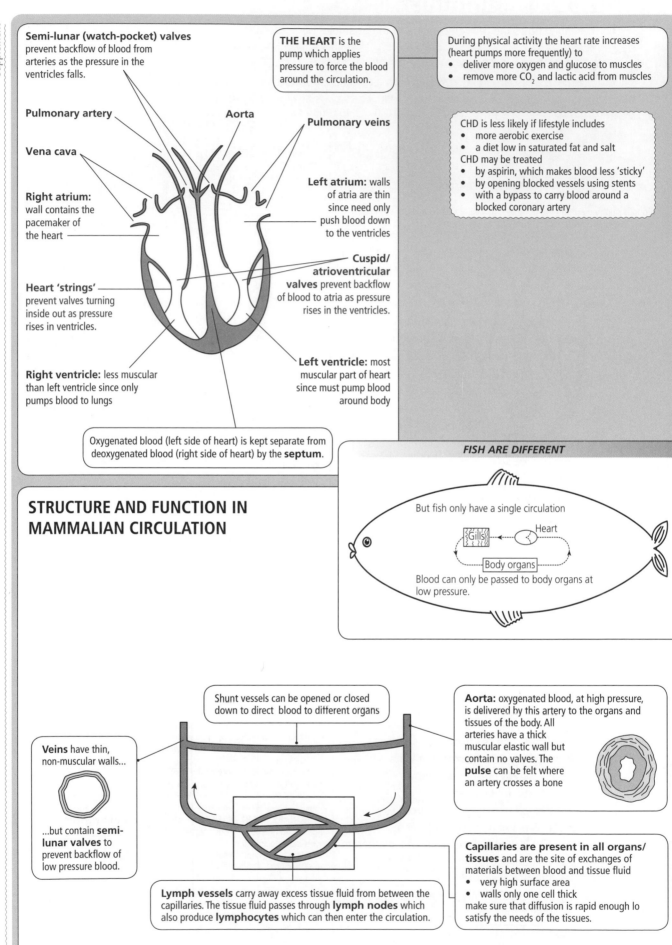

Semi-lunar (watch-pocket) valves prevent backflow of blood from arteries as the pressure in the ventricles falls.

THE HEART is the pump which applies pressure to force the blood around the circulation.

During physical activity the heart rate increases (heart pumps more frequently) to
- deliver more oxygen and glucose to muscles
- remove more CO_2 and lactic acid from muscles

Pulmonary artery

Aorta

Pulmonary veins

Vena cava

CHD is less likely if lifestyle includes
- more aerobic exercise
- a diet low in saturated fat and salt

CHD may be treated
- by aspirin, which makes blood less 'sticky'
- by opening blocked vessels using stents
- with a bypass to carry blood around a blocked coronary artery

Left atrium: walls of atria are thin since need only push blood down to the ventricles

Right atrium: wall contains the pacemaker of the heart

Cuspid/ atrioventricular valves prevent backflow of blood to atria as pressure rises in the ventricles.

Heart 'strings' prevent valves turning inside out as pressure rises in ventricles.

Left ventricle: most muscular part of heart since must pump blood around body

Right ventricle: less muscular than left ventricle since only pumps blood to lungs

Oxygenated blood (left side of heart) is kept separate from deoxygenated blood (right side of heart) by the **septum**.

STRUCTURE AND FUNCTION IN MAMMALIAN CIRCULATION

FISH ARE DIFFERENT

But fish only have a single circulation

Gills → Heart

Body organs

Blood can only be passed to body organs at low pressure.

Shunt vessels can be opened or closed down to direct blood to different organs

Aorta: oxygenated blood, at high pressure, is delivered by this artery to the organs and tissues of the body. All arteries have a thick muscular elastic wall but contain no valves. The **pulse** can be felt where an artery crosses a bone

Veins have thin, non-muscular walls...

...but contain **semi-lunar valves** to prevent backflow of low pressure blood.

Lymph vessels carry away excess tissue fluid from between the capillaries. The tissue fluid passes through **lymph nodes** which also produce **lymphocytes** which can then enter the circulation.

Capillaries are present in all organs/ tissues and are the site of exchanges of materials between blood and tissue fluid
- very high surface area
- walls only one cell thick

make sure that diffusion is rapid enough to satisfy the needs of the tissues.

1. The table shows the cell composition of three samples of blood.

	JACK	JAMES	JULIAN
Red cells / number per mm³	8 200 000	5 000 000	2 200 000
White cells / number per mm³	500	8000	5000
Platelets / number per mm³	280 000	255 000	1000

a. Which person is most likely to have recently lived at high altitude? Explain your answer. (2)

b. Which person would be least likely to resist infection by a virus? Explain your answer. (2)

c. People at risk from heart disease are sometimes recommended to take half an aspirin each day. Aspirin reduces the clotting of blood, and helps to prevent blood clots blocking narrowed arteries. Which person is most likely to be taking this drug each day? Explain your answer. (2)

d. Iron deficiency in the diet can cause a condition called anaemia. Which person is likely to show the symptoms of anaemia? Explain your answer. (2)

e. These three samples were all taken from 25-year-old men. Explain why this makes comparisons between them more valid. (2)

2. The diagram below shows three types of blood vessel in the human circulation.

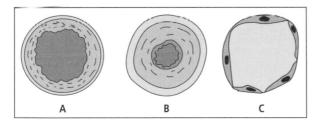

Use the label letters (A, B or C) to identify the vessels that

a. have walls one cell thick

b. allow amino acids to cross the walls

c. have valves to prevent back-flow

d. carry blood to the heart

e. carry blood under high pressure

f. pulse as blood flows through them

g. can be blocked by fatty tissue called atheroma

h. carry blood away from the heart

i. have thick elastic walls

j. increase greatly in number after long periods of athletic training (10)

3. Examine this bar chart.

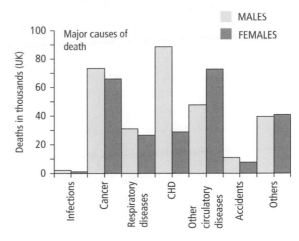

a. How many deaths in the UK result from CHD? (1)

b. How many deaths result from CHD and other circulatory diseases combined? (2)

c. What proportion of all deaths in the UK is due to CHD? (2)

d. Redraw the graph in the form of a pie chart. Which of the two – bar chart or pie chart – gives the information from **(c)** most clearly? Explain your answer. (5)

4. The diagram below shows part of the human circulation.

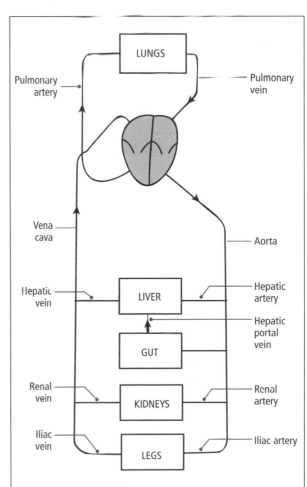

a. Name the blood vessels and chambers of the heart through which
   i. a molecule of amino acid absorbed by the gut passes on its way to build up muscle in the heart (3)
   ii. a red blood cell passes as it delivers oxygen from the lungs to the liver and returns to be reoxygenated. (3)
b. A poster in a doctor's surgery had the headline:

A HEALTHY LIFESTYLE MEANS A HEALTHY HEART

   i. Suggest **two** features of a person's lifestyle that would give them a good chance of having a healthy heart. (2)
   ii. Suggest **two different** features of a person's lifestyle that might lead to heart disease. (2)

Supplement

5. The graph below shows pressure changes, measured in millimetres of mercury, that occur during one complete beat of a human heart. **Data for pressures in the right atrium and the right ventricle are not shown**.
Look at the graph and answer the following questions.

a. i. How long does one complete beat of the heart last? (1)
   ii. Calculate the number of beats per minute. Show your working. (2)
b. i. The maximum pressure reached by the left ventricle is five times that reached by the right ventricle.
      Using the data in the graph calculate the maximum pressure reached in the right ventricle.
      Show your working. (1)
   ii. From your knowledge of the structure of the heart, explain how this difference in pressure between the ventricles is achieved. (1)
c. Explain the connection between the pressure in the left ventricle and the pressure in the aorta between 0.3 and 0.5 seconds. (1)
d. From your knowledge of how the heart works, suggest which valves close at points X and Y. In each case explain why the valve closes at that point.
   Name of valve closing at X: ....
   Reason for closing: ....
   Name of valve closing at Y: ....
   Reason for closing: .... (4)

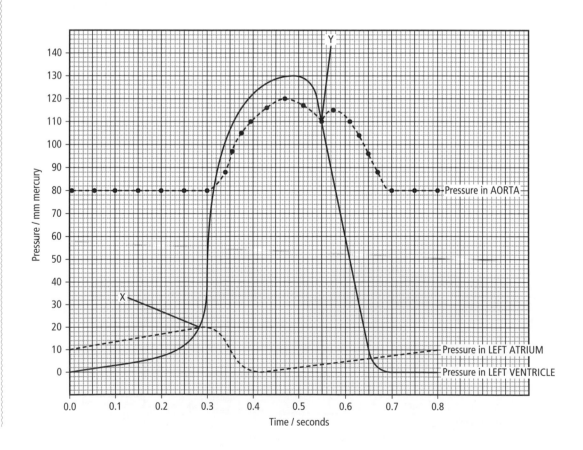

6. The diagram below shows an external view of the heart and its blood vessels.

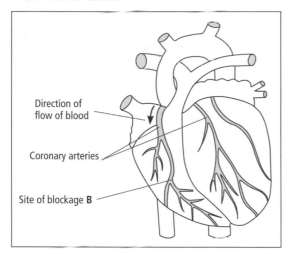

Direction of flow of blood

Coronary arteries

Site of blockage **B**

a.   The coronary arteries supply heart tissue with useful substances. Coronary veins remove waste substances.
   i.   Name **two** useful substances the coronary arteries will supply.   (2)
   ii   Name **one** waste substance the coronary veins will remove.   (1)

b.   The tissue forming the wall of the left ventricle responds when it is stimulated by electrical impulses.
   i.   Name this type of tissue.   (1)
   ii. Describe how this tissue will respond when stimulated.   (1)
   iii. Describe the effect of this response on the contents of the left ventricle.   (2)

c.   The coronary arteries can become blocked with a fatty deposit, leading to a heart attack.
   i.   State two likely causes of this type of blockage.   (2)
   ii. A blockage occurs at point **B** in the coronary artery. On the diagram, shade in the parts of the artery affected by this blockage.   (1)

d.   Veins have different structures from arteries. State two features of veins and explain how these features enable them to function efficiently.   (4)

Cambridge IGCSE Biology 0610 Paper 3 Q3 November 2005

TRANSPORT IN ANIMALS: Crossword

ACROSS:
1   Lipid which may block blood vessels
3   Controls the rate of heartbeat – sometimes artificially
4   Tissue that contracts as the heart beats
7   Can be raised by stress or excitement
9   Vessels that carry blood away from the heart
10 Artery supplying the heart muscle with oxygen and glucose
12 With 2 down – prevents blood flowing back into heart
15 Strong tissue that prevents valves turning inside out
16 With 18 across – pumping chamber of the heart forcing blood to the body (4, 9)
18 See 16 across
19 Circulation for the lungs
21 With 20 down – receiving chamber for blood from the body
22 Vessels carrying blood back towards the heart

DOWN:
1   Tiny blood vessels for exchange of materials between blood and tissues
2   Structure that ensures one way blood flow
5   This can increase the risk of heart disease, as well as damaging the lungs
6   Main vein of the body
8   Benefits the heart, but too much can be dangerous
11 Blood pressure as heart relaxes
13 Blood pressure as heart contracts
14 The main artery
17 Operation to get past a blocked coronary artery
19 Count of the beats of the heart
20 See 21 across

Chapter 16:
Defence against disease

Natural defence systems of the body
prevent infection and disease

PATHOGENS CAUSE DISEASE. Many organisms may colonize the body of a human – the body is **warm, moist** and a **good food source**. These organisms may compete with human cells for nutrients or may produce by-products that are poisonous to human cells. This will affect the normal function of the cells, i.e. it will cause 'disease'. Pathogens include viruses and bacteria.

TYPE OF PATHOGEN	DISEASE	SYMPTOMS
VIRUS Protein coat Nucleic acid	AIDS	Damage to immune system so risk of infections such as pneumonia. **Treat with drugs which stop replication of the virus, but antibiotics have no effect on the virus.**
BACTERIUM Slime coat Cell wall 'Naked' DNA (not in chromosome)	Cholera	Diarrhoea (production of watery faeces) causes dehydration and loss of salts from body. **Control with antibiotics and treat by oral rehydration therapy.**

A **transmissable** disease is one in which the pathogen may be passed from one host to another. This may be DIRECT (e.g. through blood) or INDIRECT (e.g. from contaminated food or from animals).

PHAGOCYTES DESTROY PATHOGENS BY INGESTING THEM

Phagocytes are large white blood cells. They are 'attracted to' wounds or sites of infection by chemical messages. They leave the blood vessels and destroy any pathogens they recognize.

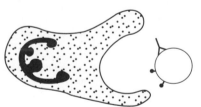

A pathogen is recognized by its surface proteins or by antibodies that 'label' it as dangerous.

Pathogens are destroyed by digestive enzymes secreted into 'food sac'.

LYMPHOCYTES PRODUCE PROTECTIVE ANTIBODIES

Lymphocytes are white blood cells that are found in the blood and in lymph nodes (swellings in the lymphatic system). They are stimulated by the presence of pathogens to manufacture and release special proteins called **antibodies**, which can recognize, bind to and help to destroy pathogens.
The body can be triggered to produce the correct antibodies by vaccination (**active immunity**). Protection can also result from a supply of ready made antibodies (e.g. in mother's milk) (**passive immunity**).

This end of the antibody acts as a signal to phagocytes to remove this pathogen

The forked end of the antibody 'recognizes' surface protein of pathogen

Surface protein identifies pathogen as non-human cells

SKIN IS THE FIRST BARRIER: the outer layer of the skin, the **epidermis**, is waxy and impermeable to water and to pathogens (although many microorganisms can live on its surface). Where there are natural 'gaps' in the skin there may be protective secretions to prevent entry of pathogens, for example:

ORIFICE	FUNCTION	PROTECTED BY
Mouth	Entry of food	Hydrochloric acid in stomach
Eyes	Entry of light	Lysozyme in tears
Ears	Entry of sound	Bactericidal ('bacteria-killing') wax

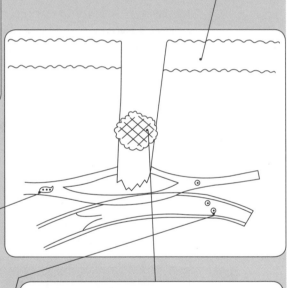

BLOOD CLOTTING PLUGS WOUNDS: a natural defence, largely due to **blood proteins** and **platelets**. It is able to block any unnatural gaps in the skin.
• prevents excessive blood loss
• prevents entry of pathogens

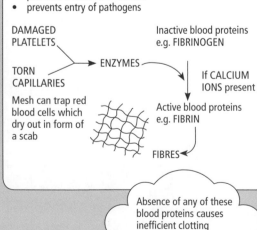

DAMAGED PLATELETS

TORN CAPILLARIES

Mesh can trap red blood cells which dry out in form of a scab

Inactive blood proteins e.g. FIBRINOGEN

ENZYMES

If CALCIUM IONS present

Active blood proteins e.g. FIBRIN

FIBRES

Absence of any of these blood proteins causes inefficient clotting → severe bleeding: **haemophilia**.

SEXUALLY TRANSMITTED INFECTIONS (STIs)
- are transmitted via body fluids through sexual contact
- **gonorrhoea** is a bacterial infection so may be controlled with antibiotics
- **AIDS** may be caused by infection with **HIV** (Human Immunodeficiency Virus)
- AIDS cannot be controlled by antibiotics
- HIV reduces human lymphocyte numbers so fewer antibodies can be produced. This may result in infection by other pathogens.

The risk of both AIDS and gonorrhoea can be reduced by not 'mixing' body fluids
- wear a condom during intercourse
- do not share needles for intravenous injections

FIGHTING DISEASE: Several different strategies are important
- **hygienic food preparation** – to kill microbes (eg Pasteurisation of milk), to remove them from kitchen surfaces (using antiseptics) and preventing microbes from reproducing (refrigeration and freezing).
- **personal hygiene** – washing hands and brushing teeth, as well as regular health checks
- **sewage treatment** – using a mixture of biological (decomposition of wastes) and chemical (sterilisation with chlorine) methods to remove pathogens from water supplies
- **waste disposal** – removes possible food sources for microbes from contact with humans

VACCINATION can enhance the body's natural defences against disease

THE IMMUNE RESPONSE MAY BE ENHANCED. Immunity is the body's use of antibodies (to combat invasion by pathogens).

Immunity may be:

ACTIVE: the individual is provoked to make his or her own antibodies

NATURAL
Pathogen infects individual

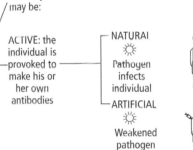

Individual contracts disease but survives, makes antibodies and is now immune to further infection by the same pathogen e.g. immunity develops to different strains of the common cold.

ARTIFICIAL
Weakened pathogen (vaccine)

Injection of vaccine does not cause disease but lymphocytes do produce antibodies individual now immune to this pathogen e.g. vaccination against rubella virus (causes German measles) in babies.

PASSIVE: the individual is protected by a supply of pre-formed antibodies

NATURAL

Mother's antibodies can cross the placenta and are delivered in breast milk: newborn child will be temporarily immune to pathogens for which mother produced antibodies.

ARTIFICIAL

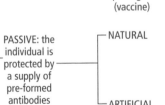

Antibodies collected from blood of laboratory animal e.g. horse or rabbit.

INJECTION

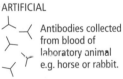

Adult is now immune to disease which is too fast-acting for own immune system to deal with e.g. injection of anti-tetanus vaccine following a deep, dirty cut or wound. NB This offers only a TEMPORARY immunity. Memory cells are not produced in passive immunity.

VACCINE PRODUCTION
Use either
- Dead pathogens e.g. Whooping cough vaccine.
- Weakened pathogens e.g. Oral polio vaccine.
- Genetically engineered fragments the proteins from the pathogens surface which are recognized by lymphocytes e.g. Hepatitis B viral coat protein

THE BEST WAYS TO AVOID DISEASE

- good, balanced diet
- not smoking
- regular exercise
- controlling intake of alcohol and other drugs
- personal hygiene, including sensible sexual behavior

LIVE WELL

1. The diagram below shows the stages involved in one process of defence against disease.

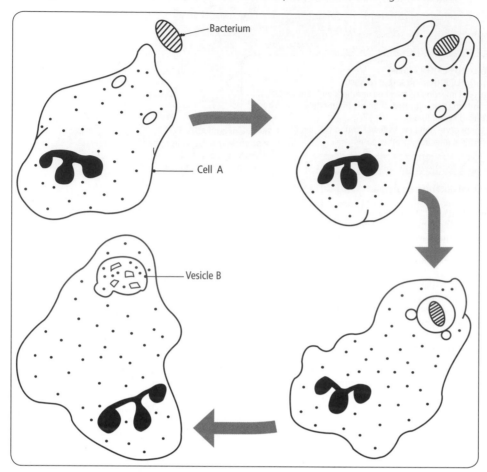

Bacterium

Cell A

Vesicle B

a. Name the cell labelled **A**. (1)
b. Name the process that the diagrams represent. (1)
c. How does cell **A** recognize the bacterium as a 'foreign' organism? (1)
d. What happens inside vesicle **B** to destroy the bacterium? (2)
e. Name **two** diseases in humans that are caused by bacteria. (2)
f. What help can a doctor offer in the defence against bacterial diseases? (1)

2. The graph shows the levels of antibody in the blood following two injections of a vaccine.

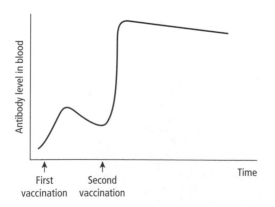

Antibody level in blood

Time

First
vaccination

Second
vaccination

a. What is meant by the term *vaccine*? (2)
b. What is an antibody? (1)
c. How does the response to the second injection differ from the response to the first injection? (3)

d. What is the advantage to a person in receiving the vaccine in two doses rather than as a single injection? (1)
e. This response of the body is an example of immunity. There are different ways of becoming immune, but each way is either 'active' or 'passive'. Complete the table below to show whether each of the methods described is active or passive. (5)

EXAMPLE	METHOD OF BECOMING IMMUNE	ACTIVE OR PASSIVE IMMUNITY
A	Receiving a vaccine	
B	Becoming infected with a virus	
C	Receiving a serum	
D	A baby feeding on colostrum	
E	A fetus receiving antibodies across the placenta	

3. a. The diseases AIDS, athlete's foot, malaria and cholera are caused by a number of different organisms.
   i.   Match up each of these diseases with the type of organism that causes it.

      A   AIDS   B   cholera
      C   athlete's foot   D   malaria
      1   bacterium   2   protoctistan
      3   virus   4   fungus   (4)
   ii. Give **one** way in which each of the diseases is spread. (4)

  b. The effectiveness of different antibiotics can be investigated using cultures of bacteria grown on nutrient agar in Petri dishes. The nutrient agar is heated to 120 °C in an autoclave, and then poured into sterile Petri dishes and left until it hardens. The investigation, and the results, are shown in the diagram below:

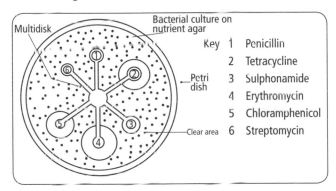

   i.   The nutrient agar contains sugar, amino acids, vitamins and minerals. Why? (1)
   ii. Why is the nutrient agar pre-heated to a high temperature? (1)
   iii. Which is the most effective antibiotic among those used in this investigation? Explain how you arrived at your answer. (2)
   iv. Name **two** diseases that could not be treated with antibiotics. (2)

4. Over a period of ten years, an antibiotic was used in a hospital to treat an infection. The figure below shows the amount of antibiotic used and the proportion of bacteria that survived treatment with the antibiotic over this period of time.

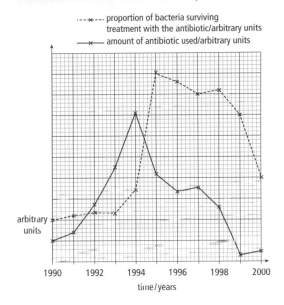

  a. Name an antibiotic. (1)
  b. State the period of time during which the antibiotic was most effective at treating the infection in the hospital. (1)
  c. Suggest and explain possible causes for the increase in the proportion of bacteria that survived treatment with the antibiotic after 1994. (5)
  d. Suggest
   i.   two reasons for the decreased use of this antibiotic after 1997, (2)
   ii. two possible ways of controlling the infection in the hospital after 1997. (2)

   Cambridge O Level Biology 5090 Paper 2 Q5 June 2009

REVISION SUMMARY: Fill in the missing words

Complete the passage below, using terms from the list. You may use each term once, more than once or not at all.

DISEASE, PATHOGENS, FIBRINOGEN, PLATELETS, WHITE BLOOD CELLS, PROTEINS, BLOOD LOSS, RED BLOOD CELLS, HAEMOPHILIA, FIBRIN

Damage to tissues results in the formation of clots. These have two basic functions – to reduce .... and to prevent the entry of ...., which could cause ....
    Temporary clots form when the .... stick to each other, but more permanent wounds need stronger barriers. The formation of more permanent clots involves several factors that are soluble .... in the blood. A series of reactions eventually causes soluble .... to be converted to insoluble ...., which forms a mesh. This mesh traps .... to form a strong barrier.
    The absence of a clotting factor may lead to a disease called ....

Breathing, gaseous exchange and respiration

Oxygen gas is needed for respiration in humans. Carbon dioxide gas is a waste product of respiration. Breathing takes gases into and out of the body.

BREATHING: the action of drawing air into the body (inhaling) and pushing air and waste gases out (exhaling).

Breathing in:
- external intercostal muscles between the ribs contract, causing the chest to expand
- diaphragm flattens, increasing the volume of the chest and reducing gas pressure in the lungs
- air enters the lungs
- **inhaled air** has about 21 per cent **oxygen** and 0.03 per cent carbon dioxide

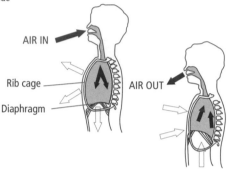

AIR IN

Rib cage

AIR OUT

Diaphragm

Breathing out:
- external intercostal muscles between the ribs relax
- diaphragm raised up, decreasing the volume of the chest cavity and increasing gas pressure in the lungs
- air 'pushed' out of the lungs
- **exhaled air** has about 4 per cent **carbon dioxide**, 17 per cent oxygen, and is saturated with water vapour
- this carbon dioxide can be detected because it turns **lime water** milky: the carbon dioxide reacts with the calcium ions in the lime water to form small particles of chalk. The CO_2 will also turn **hydrogen carbonate indicator** from red to yellow-orange.

CHANGES BETWEEN INSPIRED AND EXPIRED AIR

Oxygen	: falls as O_2 removed for aerobic respiration
Carbon dioxide	: increases as CO_2 produced by aerobic respiration
Nitrogen	: almost no change as N_2 is neither used nor produced by animal cells
Water vapour	: increases as water evaporates from the lining of the air sacs

PROTECTING THE AIRWAYS

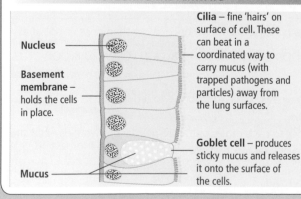

Nucleus

Basement membrane – holds the cells in place.

Mucus

Cilia – fine 'hairs' on surface of cell. These can beat in a coordinated way to carry mucus (with trapped pathogens and particles) away from the lung surfaces.

Goblet cell – produces sticky mucus and releases it onto the surface of the cells.

GASEOUS EXCHANGE: the movement of oxygen from inhaled air into the blood and the movement of carbon dioxide from the blood into the airways of the lungs.

The airways of the lungs get smaller and smaller as they divide. The **bronchioles** end in thousands of tiny air sacs called **alveoli**. These have thin, moist walls so that gases can pass in and out easily.

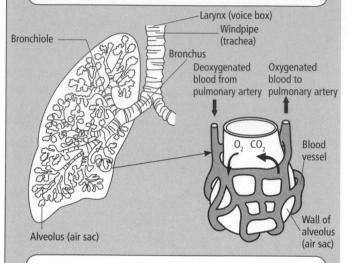

Bronchiole

Larynx (voice box)

Windpipe (trachea)

Bronchus

Deoxygenated blood from pulmonary artery

Oxygenated blood to pulmonary artery

O_2 CO_2

Blood vessel

Wall of alveolus (air sac)

Alveolus (air sac)

Each **alveolus** is covered by thin capillaries. Oxygen diffuses into the blood where it is taken up by red blood cells. Carbon dioxide diffuses from the blood into the alveolus, ready to be exhaled.

The lungs are an ideal surface for gas exchange
- large surface area
- thin surface lining the air sacs
- good blood supply
- ventilated with air by breathing movements

Physical activity
- increases both **rate** and **depth** of breathing
- to supply more O_2 and remove more CO_2
- to keep up aerobic respiration to supply energy

EXERCISE AND BREATHING
- Exercise produces more **carbon dioxide**
- Extra carbon dioxide **lowers the pH of the blood**
- Change in pH **stimulates the breathing control centre in the brain**
- Breathing control centre **increases rate and depth of breathing**
- Excess carbon dioxide is cleared from the blood

Supplement

1. Match up the two lists below to explain the ideal features of a gas exchange surface.

	FEATURE		VALUE IN GAS EXCHANGE
A	moist	a	gases do not have to diffuse very far
B	large surface area	b	oxygen is quickly removed so that diffusion can continue
C	thin	c	increase the number of molecules that can diffuse across at the same time
D	well ventilated	d	cells die if they dry out
E	good blood supply	e	regular supply of fresh air keeps up the concentration gradients for oxygen and carbon dioxide

(5)

2. Complete the following paragraphs. The words in the following list may be used once, more than once or not at all.

GILLS, OXYGEN, ENERGY, THIN, AMOEBA, NITROGEN, MOIST, SPIRACLES, RESPIRATION, EXCRETION, SURFACE AREA, BLOOD, TRACHEOLES, VENTILATION, CARBON DIOXIDE

All living organisms require......................, which is released from the process of .... The most efficient form of this process requires the gas ....and produces the waste gas........................To keep this energy–releasing process going the organism must have a gas exchange surface – this surface has certain properties, it has a large........................, a........................ membrane so that diffusion distances are short and a........................ layer (since cells die if they dry out). In addition the most advanced systems have a means of........................ to move the gases over the surface, and are close to a........................supply to transport gases between the surface and the living tissues.

In simple organisms such as........................enough oxygen can diffuse through the outer cell membrane, but larger organisms require more oxygen than diffusion alone can supply. Insects have small 'holes', called........................, in their body covering – these connect to tubes called........................ that lead directly to the working tissues. More active and larger animals may have a specialized gas exchange surface, such as the........................found in fish. (13)

3. a. The diagram shows the position of the diaphragm and ribs at rest and after inhaling.

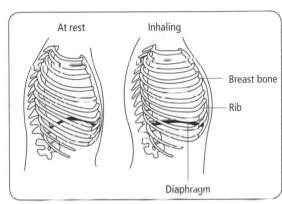

Explain how movements of the diaphragm and ribs cause air to enter the lungs during inhalation. (4)

b. A scientist investigated the effect of increasing carbon dioxide concentration on the breathing rate and on the total volume of air breathed per minute. The results are shown in the graph below.

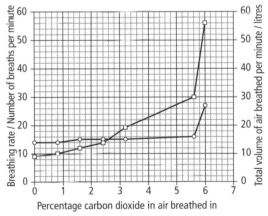

-○- Breathing rate
-□- Total volume of air breathed per minute/litres

i. Use the graph to describe the effect of increasing the carbon dioxide concentration on the total volume of air breathed per minute. (2)

ii. Suggest why the total volume of inhaled air is **not** directly proportional to the rate of breathing. (2)

4. The tables below contain information on the effects of smoking on health.

CIGARETTES SMOKED PER DAY	ANNUAL DEATH RATE PER 1000 MEN FROM LUNG CANCER
0	0.1
5	0.4
10	0.6
15	1.0
20	1.5
25	1.7
30	2.0
35	2.4
40	2.7
45	3.0
50	3.6

PERIOD SINCE GIVING UP SMOKING / YEARS	ANNUAL DEATH RATE PER 1000 MEN FROM LUNG CANCER
0	1.80
5	0.96
10	0.55
15	0.31
20	0.25
25	0.20

a. Plot these figures as two graphs. (8)
b. Out of a group of 5000 men smoking 20 cigarettes per day, how many are likely to die of lung cancer in one year? (1)
c. How long does it take to reduce the risk of lung cancer by 60 per cent by giving up smoking? Give your answer to the nearest year. (1)
d. Name the addictive substance found in tobacco smoke. (1)
e. Smokers often have blood that has lost some of its 'redness': this can be seen by the lips taking on a pale blue colour. Suggest a reason for this. (1)
f. Name two other diseases that can be caused by cigarette smoking. (2)

5. A student's breathing was monitored before and after vigorous exercise. The student breathed in and out through special apparatus. The graphs show the changes in the volume of air inside the apparatus.
Each time the student breathed in, the line on the graph dropped. Each time the student breathed out, the line went up.

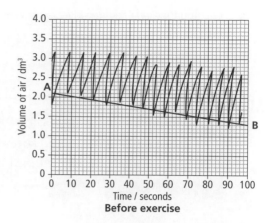
Before exercise

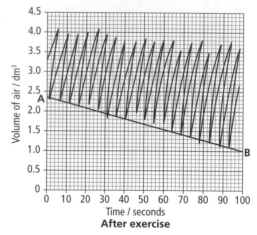
After exercise

a. How many times did the student breathe in per minute:
before exercise;
after exercise? (1)
b. On each graph, the line **A–B** shows how much oxygen was used. The rate of oxygen use before exercise was 0.5 dm³ per minute.
Calculate the rate of oxygen use after exercise.

Rate of oxygen use after exercise =
............................dm³ per minute (2)

c. The breathing rate and the amount of oxygen used were still higher after exercise, even though the student sat down to rest.
Why were they still higher? (4)

6. The figure below shows an alveolus in which gaseous exchange takes place.

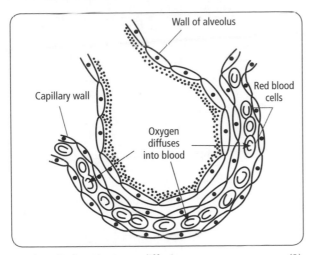

Wall of alveolus

Capillary wall

Red blood cells

Oxygen diffuses into blood

a. i. Define the term *diffusion*. (2)

ii. State what causes oxygen to diffuse into the blood from the alveoli. (1)

iii. List three features of gaseous exchange surfaces in animals, such as humans. (3)

b. i. At high altitudes there is less oxygen in the air than at sea level. Suggest how this might affect the uptake of oxygen in the alveoli. (2)

ii. In the past some athletes have cheated by injecting themselves with extra red blood cells before a major competition. Predict how this increase in red blood cells might affect their performance. (2)

Cambridge IGCSE Biology 0610 Paper 2 Q9 November 2006

Revision Summary: Fill in the missing words

Complete the following paragraph by filling in the missing words. The words in the following list may be used once, more than once or not at all.

ENERGY, HYDROGEN CARBONATE, LEFT ATRIUM, PULMONARY, RESPIRATION, THIN, OXYGEN, SURFACE AREA, RENAL, ALVEOLI, CAPILLARIES, ARTERIES, TRACHEA, RIGHT VENTRICLE, CARBON DIOXIDE, DIFFUSION.

Deoxygenated blood arrives at the lungs in the .... arteries. Oxygen has been removed from the blood by cells that are carrying out .... to release .... needed to carry out their functions. This blood also contains a relatively high concentration of the gas........................, which is carried dissolved in the plasma as .... ions. Each artery branches many times to form........................ which are well adapted to allow the exchange of gases because they are........................-walled and have a very large........................ These small vessels lie very close to the .... of the lungs, and it is here that gas exchange takes place. The gas ....moves out of the blood and the gas ....moves into the blood. Both gases move by the process of .... Oxygenated blood then leaves the lungs in the........................vein that returns blood to the heart at the chamber called the .... (14)

Chapter 18:
Respiration and the release of energy

Respiration may be aerobic or anaerobic

Respiration is the chemical process in which food, usually glucose, is oxidized to release energy. This involves the action of enzymes in cells. The energy is used to carry out metabolism (the work needed to keep organisms alive):

- mechanical work (movement of muscles),
- growth and repair (by cell division)
- chemical work (active transport and chemical building in plants and animals)
- production of heat needed for **metabolic** (life) processes, e.g. providing the optimum temperature for the action of enzymes
- the passage of nerve impulses (pumping ions in and out of nerve cells)

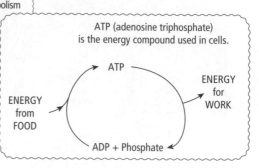

ATP (adenosine triphosphate) is the energy compound used in cells.

ENERGY from FOOD → ATP → ENERGY for WORK → ADP + Phosphate →

AEROBIC RESPIRATION

In most organisms, respiration uses **oxygen**. This is called **aerobic respiration**, and takes place in the MITOCHONDRIA.

The chemical reactions are complex but we can summarize aerobic respiration in this equation:

Glucose   Oxygen   Carbon dioxide   Water

 $C_6H_{12}O_6$ + $6O_2$ ⟹ $6CO_2$ + $6H_2O$ + energy

about 38 molecules of ATP for every molecule of glucose

ANAEROBIC RESPIRATION

In some cases, respiration takes place **without** oxygen. This **anaerobic respiration**:
- only partly breaks down the glucose
- releases less energy
- takes place in the cytoplasm

IN YEAST

Glucose ⟹ Carbon dioxide + Ethanol + Energy (2 ATP)

$C_6H_{12}O_6$ ⟹ $2C_2H_5OH + 2CO_2$

Important in BREWING and in BAKING

IN MUSCLE CELLS DURING EXERCISE

Glucose ⟹ Lactic acid + Energy (2 ATP)

After exercise oxygen is needed for recovery!

During sport or strenuous activity, muscle cells can run out of oxygen. They then start to respire anaerobically. The lactic acid produced builds up in the muscles and can cause painful 'cramps' until you have taken in enough oxygen to break down the lactic acid and repay the **oxygen debt**.

1. Respiration is a feature of living organisms. Make a list of **five** other characteristics of living organisms. (5)

2. The diagram shows how much energy is required for various activities. The figures are given in kilojoules (kJ) per hour.

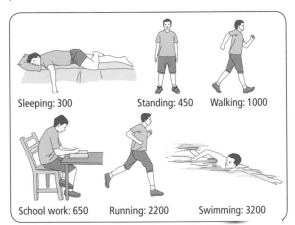

Sleeping: 300   Standing: 450   Walking: 1000

School work: 650   Running: 2200   Swimming: 3200

   a. Plot this information on a bar chart. (4)
   b. During a typical school day Sam stood for 1 hour, walked for 1 hour, ran for half an hour and did school work for 4 hours. How much energy did he use? (2)
   c. Sleeping uses 300 kJ in one hour. Why is energy needed during periods of sleeping? (2)
   d. Suggest why swimming uses more energy than running. (2)

3. a. A student investigated the energy content of a seed.

   A seed was weighed and its mass recorded in the table opposite. The seed was firmly attached to the end of a mounted needle. A large test tube containing 20 cm³ of water was held in a clamp stand, with a thermometer and a stirrer. The apparatus is shown below.

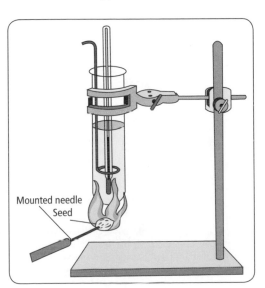

Mounted needle
Seed

   • The temperature of the water at the start was recorded.
   • The seed was set alight by placing it in a flame for a few seconds.
   • The burning seed was held under the test tube until the seed was completely burnt.
   • The water was stirred immediately. The highest temperature of the water was recorded.
   i. Complete the table by calculating the rise in temperature. (1)

MASS OF SEED / g	0.5
VOLUME OF WATER / cm³	20
TEMPERATURE AT THE START / °C	29
HIGHEST TEMPERATURE / °C	79
RISE IN TEMPERATURE / °C	

The energy contained in the seed can be calculated using the formula below.

$$\text{energy} = \frac{\text{volume of water} \times \text{rise in temperature} \times 4.2}{\text{mass of seed} \times 1000}$$

   ii. Using the formula calculate the energy content of the seed.
   Show your working.
   Energy content ....kJg⁻¹ (2)

The same method was used to find the energy content of some food substances. The results are shown in the next table.

FOOD SUBSTANCE	Starch	Sugar	Fat	Protein
MASS OF FOOD BURNT / g	0.62	0.54	0.56	0.40
STARTING TEMPERATURE / °C	31	30	30	31
FINAL TEMPERATURE / °C	65	59	90	52
RISE IN TEMPERATURE / °C	34	29	60	21
ENERGY CONTENT / kJg⁻¹	4.61	4.51	9.00	4.41

   iii. Plot a suitable graph to compare the energy content per gram of the four different food substances **and** the seed from (a)(ii). (4)
   iv. Use this information to suggest the main food substance present in the seed. (1)
   b. Describe how you would test for the presence of reducing sugars in a seed. (3)

Cambridge IGCSE Biology 0610 Paper 6 Q1 June 2006

4. The diagrams below show four stages of an
investigation in which two mice were each fed
with glucose solution. One of the mice was fed
with a glucose solution containing radioactive
carbon (^{14}C). The other mouse was fed with a normal
(non-radioactive) glucose solution. Mice fed on glucose
solution containing low levels of radioactive carbon (^{14}C)
for short periods of time suffer no ill effects. Both mice
were approximately the same size.

a. i.   Name the gas which was responsible for
         turning the limewater milky at stage 2.   (1)

    ii. Name the living process in the cells which
         produces this gas.

    iii. This process releases energy. State **three** uses
         of this energy in mice.   (3)

b. At stage 2 the air entering the bell jar containing
    the mouse contained no carbon dioxide. Describe
    a method by which the carbon dioxide could have
    been removed.   (2)

c. Explain the importance of the mice being
    approximately the same size.   (2)

d. Describe the sequence of processes that lead to
    filter paper A becoming radioactive.   (5)

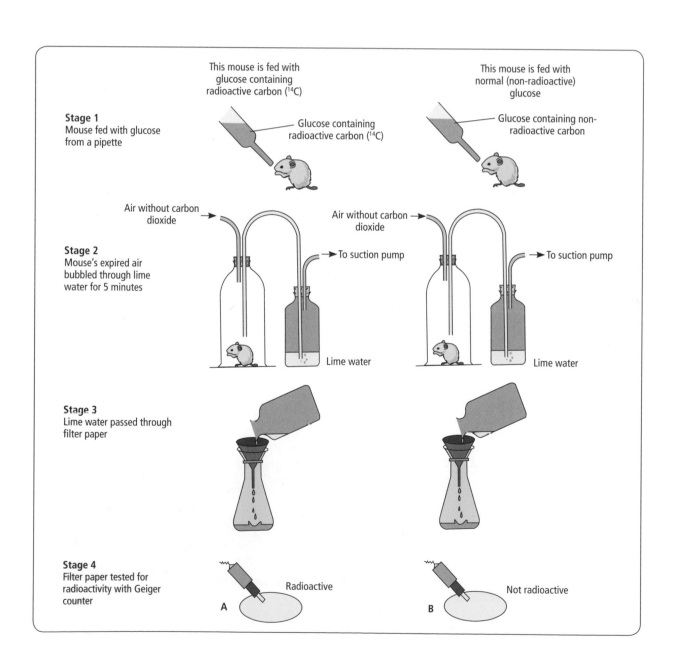

5. Here are a number of statements about aerobic respiration. State whether each of them is true (T) or false (F).

Aerobic respiration

a. uses water
b. always requires oxygen
c. occurs only in animal cells
d. releases energy
e. produces carbon dioxide
f. is controlled by enzymes
g. must have sugar as a starting material
h. releases heat
i. releases water
j. is affected by temperature
k. occurs in all living cells

(10)

REVISION SUMMARY: Fill in the missing words

Complete the following paragraph. Use terms from the list below – you may use each term once, more than once or not at all.

ANAEROBIC, MOIST, LIVING CELLS, GROWTH, ANIMALS, GLUCOSE, AEROBIC, OXYGEN, CARBON DIOXIDE, HEAT, MOVEMENT, WATER, ENERGY

Respiration is a process that occurs in all .... The raw materials are .... and, sometimes .... The purpose of the process is to release ...., and it is more efficient under .... conditions. One product of respiration can be used to carry out work; this work includes .... and ...., but because respiration is never 100 per cent efficient .... is also released and may be used to keep up body temperature in mammals and birds.

(8)

Chapter 19:
Excretion and osmoregulation

Kidney structure and function

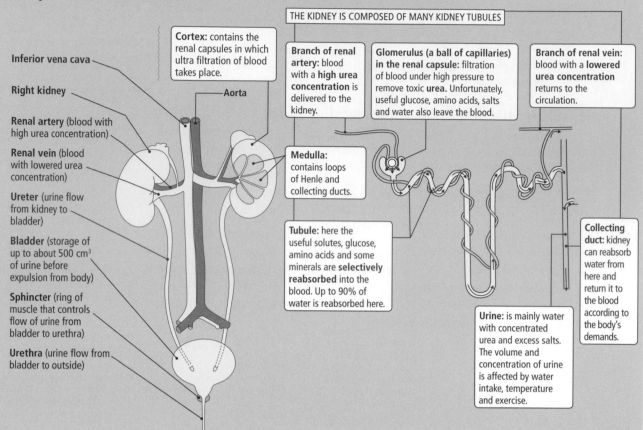

THE KIDNEY IS COMPOSED OF MANY KIDNEY TUBULES

Cortex: contains the renal capsules in which ultra filtration of blood takes place.

Branch of renal artery: blood with a **high urea concentration** is delivered to the kidney.

Glomerulus (a ball of capillaries) in the renal capsule: filtration of blood under high pressure to remove toxic **urea**. Unfortunately, useful glucose, amino acids, salts and water also leave the blood.

Branch of renal vein: blood with a **lowered urea concentration** returns to the circulation.

Inferior vena cava

Right kidney

Aorta

Renal artery (blood with high urea concentration)

Renal vein (blood with lowered urea concentration)

Ureter (urine flow from kidney to bladder)

Bladder (storage of up to about 500 cm³ of urine before expulsion from body)

Sphincter (ring of muscle that controls flow of urine from bladder to urethra)

Urethra (urine flow from bladder to outside)

Medulla: contains loops of Henle and collecting ducts.

Tubule: here the useful solutes, glucose, amino acids and some minerals are **selectively reabsorbed** into the blood. Up to 90% of water is reabsorbed here.

Collecting duct: kidney can reabsorb water from here and return it to the blood according to the body's demands.

Urine: is mainly water with concentrated urea and excess salts. The volume and concentration of urine is affected by water intake, temperature and exercise.

Excretion is removal from the body of toxic materials, the waste products of metabolism.

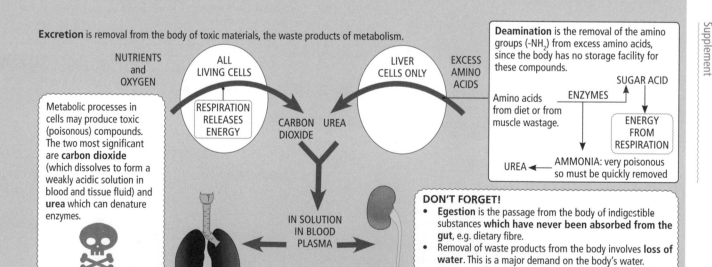

Deamination is the removal of the amino groups ($-NH_2$) from excess amino acids, since the body has no storage facility for these compounds.

NUTRIENTS and OXYGEN

ALL LIVING CELLS

RESPIRATION RELEASES ENERGY

CARBON DIOXIDE

UREA

LIVER CELLS ONLY

EXCESS AMINO ACIDS

Amino acids from diet or from muscle wastage.

ENZYMES

SUGAR ACID

ENERGY FROM RESPIRATION

AMMONIA: very poisonous so must be quickly removed

UREA

Metabolic processes in cells may produce toxic (poisonous) compounds. The two most significant are **carbon dioxide** (which dissolves to form a weakly acidic solution in blood and tissue fluid) and **urea** which can denature enzymes.

IN SOLUTION IN BLOOD PLASMA

Carbon dioxide is removed from the body by diffusion across the lung surface.

Urea is removed by ultrafiltration of the blood in the kidney, and expelled from the body in the urine.

DON'T FORGET!
- **Egestion** is the passage from the body of indigestible substances **which have never been absorbed from the gut**, e.g. dietary fibre.
- Removal of waste products from the body involves **loss of water**. This is a major demand on the body's water.
- Urea is **produced in the liver** but **excreted from the kidney**.
- Metabolic processes also produce **heat**. In excess this is dangerous (causes enzyme denaturation) and is largely removed through the lungs and the skin.

Kidney failure

If one or both kidneys fail then dialysis is used or a transplant performed to keep urea and solute concentration in the blood constant.

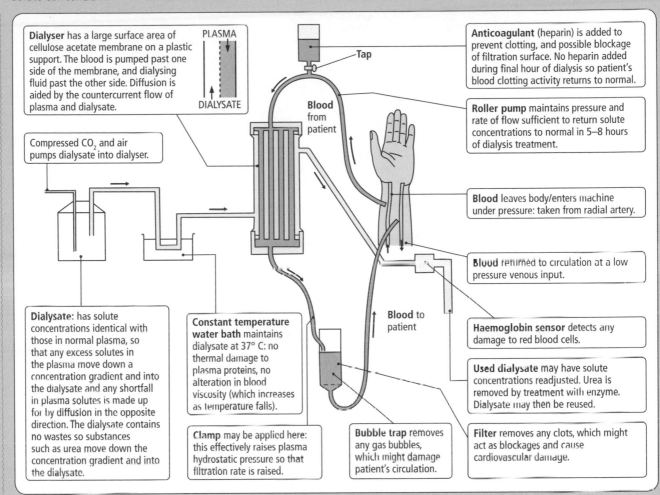

Dialyser has a large surface area of cellulose acetate membrane on a plastic support. The blood is pumped past one side of the membrane, and dialysing fluid past the other side. Diffusion is aided by the countercurrent flow of plasma and dialysate.

PLASMA

DIALYSATE

Compressed CO_2 and air pumps dialysate into dialyser.

Blood from patient

Tap

Anticoagulant (heparin) is added to prevent clotting, and possible blockage of filtration surface. No heparin added during final hour of dialysis so patient's blood clotting activity returns to normal.

Roller pump maintains pressure and rate of flow sufficient to return solute concentrations to normal in 5–8 hours of dialysis treatment.

Blood leaves body/enters machine under pressure: taken from radial artery.

Blood returned to circulation at a low pressure venous input.

Dialysate: has solute concentrations identical with those in normal plasma, so that any excess solutes in the plasma move down a concentration gradient and into the dialysate and any shortfall in plasma solutes is made up for by diffusion in the opposite direction. The dialysate contains no wastes so substances such as urea move down the concentration gradient and into the dialysate.

Constant temperature water bath maintains dialysate at 37° C: no thermal damage to plasma proteins, no alteration in blood viscosity (which increases as temperature falls).

Blood to patient

Haemoglobin sensor detects any damage to red blood cells.

Used dialysate may have solute concentrations readjusted. Urea is removed by treatment with enzyme. Dialysate may then be reused.

Clamp may be applied here: this effectively raises plasma hydrostatic pressure so that filtration rate is raised.

Bubble trap removes any gas bubbles, which might damage patient's circulation.

Filter removes any clots, which might act as blockages and cause cardiovascular damage.

KIDNEY TRANSPLANTATION

Transplantation may be necessary as renal dialysis is inconvenient for the patient and costly.

Kidney transplants have a high success rate because:

1. The vascular connections are simple.

2. Live donors may be used, so very close blood group matching is possible.

3. Because of 2 there are fewer immunosuppression-related problems in which the body's immune system reacts against the new kidney.

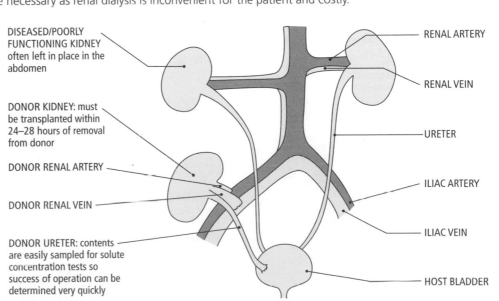

DISEASED/POORLY FUNCTIONING KIDNEY often left in place in the abdomen

DONOR KIDNEY: must be transplanted within 24–28 hours of removal from donor

DONOR RENAL ARTERY

DONOR RENAL VEIN

DONOR URETER: contents are easily sampled for solute concentration tests so success of operation can be determined very quickly

RENAL ARTERY

RENAL VEIN

URETER

ILIAC ARTERY

ILIAC VEIN

HOST BLADDER

1. The diagram shows water loss and uptake for an adult human.

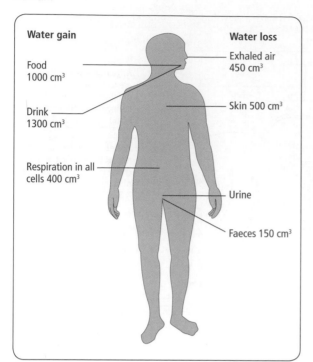

Water gain

Food 1000 cm³

Drink 1300 cm³

Respiration in all cells 400 cm³

Water loss

Exhaled air 450 cm³

Skin 500 cm³

Urine

Faeces 150 cm³

The kidneys keep the body's water content at its optimum level by controlling the water lost in the urine.

a. Use information from the diagram to calculate the mean daily loss of water in urine. Show your working. (2)

b. Describe how the kidneys control the water content of the body. Use the term 'negative feedback' in your answer. (4)

c. Sometimes kidneys fail, and the composition of the blood may change to a dangerous extent. The condition may be treated by dialysis or with a kidney transplant.

   i. Give two possible advantages of using a dialysis machine rather than having a kidney transplant.

   ii. Give two possible disadvantages of using a dialysis machine rather than having a kidney transplant. (4)

2. The diagram shows the process of filtration in the kidney.

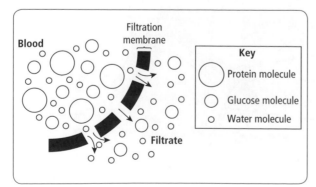

Blood

Filtration membrane

Key

◯ Protein molecule

◯ Glucose molecule

○ Water molecule

○ **Filtrate**

Use information in the diagram and your own knowledge of how the kidney works to explain why:

   i. protein molecules are not normally present in urine; (1)

   ii. glucose molecules are not normally present in urine. (3)

3. a. Define the terms

   i. excretion, (1)

   ii. egestion. (1)

   b. The kidney is an excretory organ. It produces urine that contains urea.

   i. State where in the body urea is formed. (1)

   ii. State what urea is formed from. (1)

   c. This diagram shows the urinary system and its blood supply.

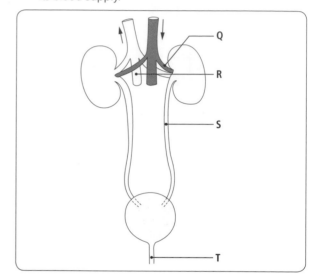

Name the parts labelled **Q**, **R**, **S** and **T**. (4)

   d. Complete the table below to show which components of the blood are also part of the urine of a healthy person.

   Use ticks (√) and crosses (X). Two boxes have already been completed. (2)

COMPONENT OF BLOOD	PRESENT IN URINE
Glucose	
Red blood cells	
Salts	
Urea	√
Water	
White blood cells	X

Cambridge IGCSE Biology 0610 Paper 2 Q3 November 2005

4. A patient has kidney failure. Every few days the patient goes into hospital for dialysis treatment.

The graph shows the concentration of urea in the patient's blood over a period of two weeks.

The concentration of urea in a healthy person's blood is shown for comparison.

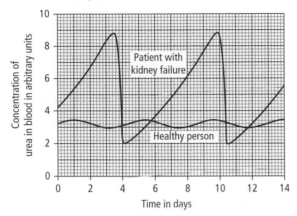

Supplement

a. Suggest why the concentration of urea in the patient's blood rises between dialysis treatments. (1)

b. If the concentration of urea in the blood rises to 9 units, the patient may die. The dialysis machine reduces the concentration of urea in the patient's blood to below the concentration in the blood of a healthy person.

Suggest **one** advantage of this to the patient. (1)

c. One alternative to treatment with a dialysis machine is to have a kidney transplant. The donor kidney is of similar tissue type to that of the patient. However, the transplanted kidney may still be rejected by the patient's immune system. State **two** other treatments which may help to prevent rejection of the transplanted kidney.

Explain your answers. (4)

EXCRETION AND OSMOREGULATION: Crossword

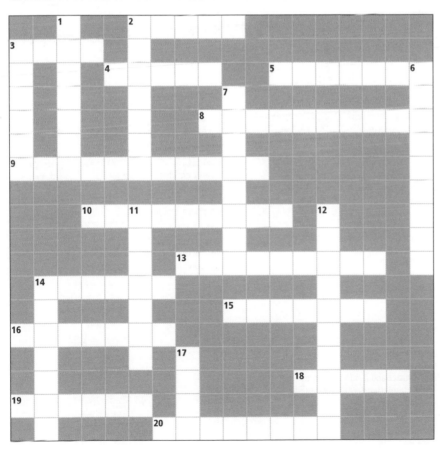

ACROSS:
2 Expelled from the body to remove urea
3 Nitrogen-containing waste product
4 This loop makes conditions which control water conservation
5 Storage sac for 3 across
8 Process that produces fluid in the kidney tubule
9 Delivers blood to the kidney (5, 6)
10 Ring of muscle that prevents embarrassing urination
13 The removal of waste materials from the body
14 Liquid part of the blood
15 Important source of energy reclaimed from fluid in the kidney tubule
16 Capsule where 8 across goes on
18 Urea is produced here
19 Urea is removed from the blood here
20 Treatment for a failing kidney

DOWN:
1 Kidney tubule
2 Tube from bladder to the environment
3 Tube from kidney to bladder
6 Carries 'purified' blood away from the kidney
7 With 11 down and 17 down: controls water conservation by the kidney
11 See 7 down
12 Deamination helps to remove these compounds if they are in excess of the body's needs
14 Solute that should never leave the plasma
17 See 7 down

The central nervous system (CNS) is formed of a brain and a spinal cord

and is connected to the various parts of the body by the peripheral nervous system. Neurones (nerve cells) carry information through the nervous system.

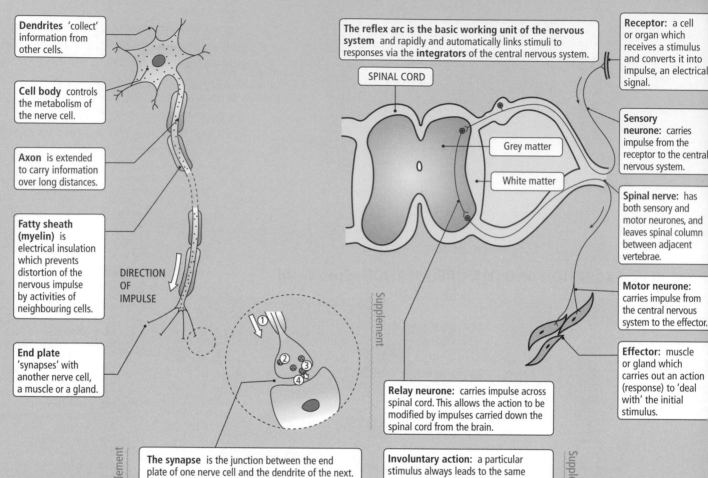

Dendrites 'collect' information from other cells.

Cell body controls the metabolism of the nerve cell.

Axon is extended to carry information over long distances.

Fatty sheath (myelin) is electrical insulation which prevents distortion of the nervous impulse by activities of neighbouring cells.

DIRECTION OF IMPULSE

End plate 'synapses' with another nerve cell, a muscle or a gland.

The reflex arc is the basic working unit of the nervous system and rapidly and automatically links stimuli to responses via the **integrators** of the central nervous system.

SPINAL CORD

Grey matter

White matter

Receptor: a cell or organ which receives a stimulus and converts it into impulse, an electrical signal.

Sensory neurone: carries impulse from the receptor to the central nervous system.

Spinal nerve: has both sensory and motor neurones, and leaves spinal column between adjacent vertebrae.

Motor neurone: carries impulse from the central nervous system to the effector.

Effector: muscle or gland which carries out an action (response) to 'deal with' the initial stimulus.

Relay neurone: carries impulse across spinal cord. This allows the action to be modified by impulses carried down the spinal cord from the brain.

The synapse is the junction between the end plate of one nerve cell and the dendrite of the next.
1. An impulse arrives at the synapse
2. At the end plate are tiny sacs containing a chemical (**neurotransmitter**)
3. The chemical is released into the synaptic gap
4. The chemical diffuses across the gap and the impulse 'restarts' on the other side.

Involuntary action: a particular stimulus always leads to the same response. There is no conscious control of the response, e.g. the the pupil reflex.

Voluntary action: requires a conscious decision by the brain. There is a choice about the response to a particular stimulus, e.g. choosing to move in or out of the shade.

Drugs may interfere with synaptic activity e.g. Amphetamine ('Speed') stimulates nervous activity because its chemical structure is similar to the neurotransmitter and it mimics it in the synaptic gap. Heroin interferes with the binding of neurotransmitter to the receptors on the next neurone, so has a 'calming' effect.

Supplement

Homeostasis: controlling body temperature involves sensory and motor neurones

Homeostasis is the process whereby the body adjusts to changes to maintain a constant internal environment.

Supplement

Many of the body's systems work best under specific chemical and physical conditions. To keep things in a steady state, the body uses a system of **detectors** and effectors. The detectors detect changes inside the body. The effectors then bring about changes in the opposite direction in order to restore equilibrium. Keeping the body's internal temperature at 37 °C is a good example of **homeostasis**.

Negative feedback: an alteration in conditions sets off a set of changes which cancel out this alteration.

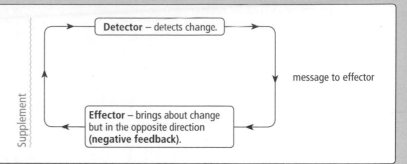

Detector – detects change.

message to effector

Effector – brings about change but in the opposite direction (**negative feedback**).

BODY TOO HOT...

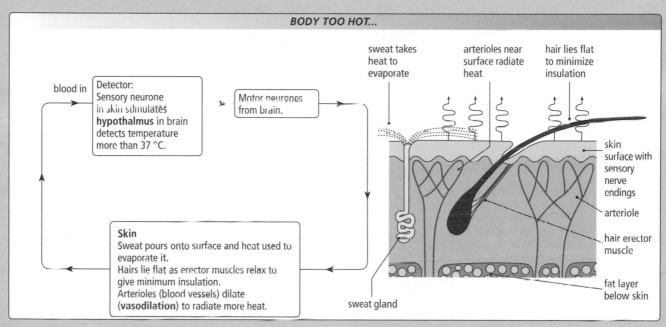

blood in

Detector:
Sensory neurone in skin stimulates **hypothalmus** in brain detects temperature more than 37 °C.

Motor neurones from brain.

sweat takes heat to evaporate

arterioles near surface radiate heat

hair lies flat to minimize insulation

Skin
Sweat pours onto surface and heat used to evaporate it.
Hairs lie flat as erector muscles relax to give minimum insulation.
Arterioles (blood vessels) dilate (**vasodilation**) to radiate more heat.

skin surface with sensory nerve endings

arteriole

hair erector muscle

fat layer below skin

sweat gland

BODY TOO COLD...

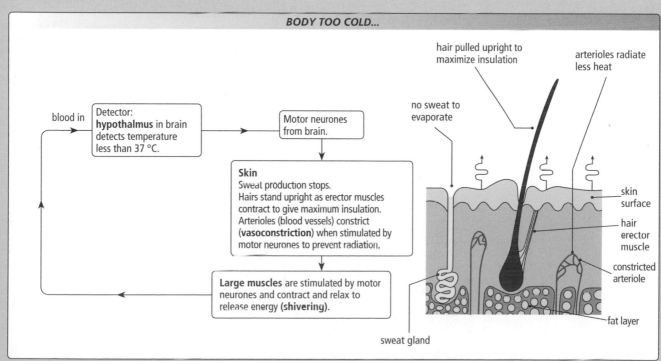

blood in

Detector:
hypothalmus in brain detects temperature less than 37 °C.

Motor neurones from brain.

hair pulled upright to maximize insulation

arterioles radiate less heat

no sweat to evaporate

Skin
Sweat production stops.
Hairs stand upright as erector muscles contract to give maximum insulation.
Arterioles (blood vessels) constrict (**vasoconstriction**) when stimulated by motor neurones to prevent radiation.

Large muscles are stimulated by motor neurones and contract and relax to release energy (**shivering**).

skin surface

hair erector muscle

constricted arteriole

fat layer

sweat gland

Drug abuse

A drug is any substance taken into the body that modifies or affects chemical reactions in the body.

ANTIBIOTICS ARE USEFUL DRUGS!
These substances are produced by *molds* and either
• kill bacteria, or
• slow down reproduction of bacteria so that the body's own immune system can destroy them.

Penicillin is a well-known example of an antibiotic. Antibiotics **do not** work against viruses.

Supplement

Antibiotics work because they interfere with the metabolism of the bacteria, e.g. stop protein synthesis. Viruses are not affected because they have no metabolism of their own – they take over the host cell's metabolism.

Supplement

Bacteria (e.g. MRSA) can become resistant to antibiotics:
• mutants in population of bacteria have resistance
• overuse of antibiotic **selects** the resistant bacteria
• resistant bacteria multiply.

HEROIN
Has its effects because it alters the way that neurotransmitters cross the synapses in the nervous system.

Causes lethargy, loss of self-control, mental and physical deterioration.

There is a serious danger of overdose, leading to coma and death.

Addiction can lead to crime as a way to obtain the drug.

Sharing needles for injection of heroin can lead to infection e.g. HIV/AIDS and hepatitis.

TOBACCO
Nicotine is addictive. Tar and smoke particles from cigarette smoke contain chemicals which cause cancer and carbon monoxide which damages the cilia in the lungs, leading to infection and breathing difficulties including **bronchitis** and **emphysema**.

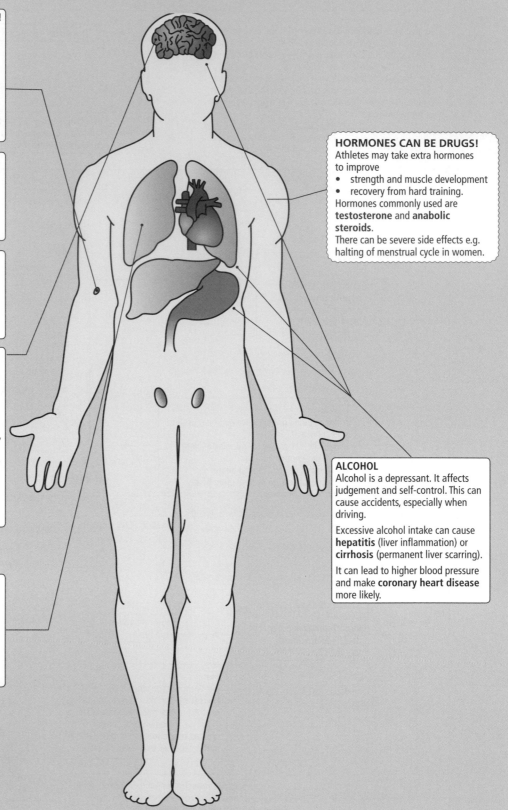

HORMONES CAN BE DRUGS!
Athletes may take extra hormones to improve
• strength and muscle development
• recovery from hard training.
Hormones commonly used are **testosterone** and **anabolic steroids**.
There can be severe side effects e.g. halting of menstrual cycle in women.

ALCOHOL
Alcohol is a depressant. It affects judgement and self-control. This can cause accidents, especially when driving.

Excessive alcohol intake can cause **hepatitis** (liver inflammation) or **cirrhosis** (permanent liver scarring).

It can lead to higher blood pressure and make **coronary heart disease** more likely.

1. The diagram shows a single neurone.

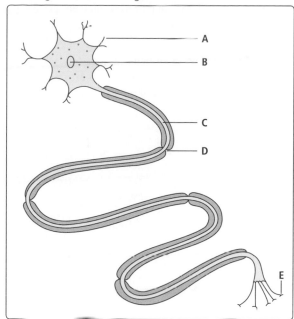

Which of the parts labelled **A** to **F**:

a.   connects with another neurone? (1)

b.   insulates the axon to limit interference with other neurones? (1)

c.   contains high concentrations of chemicals called neurotransmitters? (1)

d.   connects with an effector, such as a muscle? (1)

e.   allows a nerve impulse to 'jump' quickly along the axon? (1)

f.   contains DNA? (1)

2. The bar graph shows the units of alcohol present in some different types of alcoholic drink. The line graph relates alcohol level in the blood to the units of alcohol consumed.

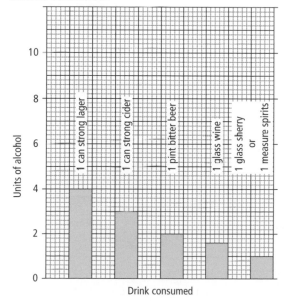

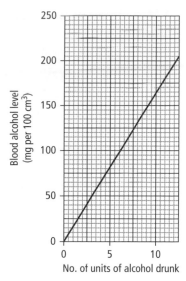

a.   How many units of alcohol are present in one can of strong lager? (1)

b.   How much alcohol would be present in the blood after drinking a pint of bitter and two measures of whisky? (2)

c.   The UK limit for blood alcohol level in a person driving a motor vehicle is 80 mg per 100 cm³. A person drank three cans of cider and one glass of wine – by how much would her blood alcohol level be over the legal limit? Show your working. (3)

d.   Why is it dangerous to drive when over the legal limit? (2)

e.   Alcohol is cleared from the bloodstream by the action of the liver, at the rate of about 1 unit per hour. How many hours after drinking the alcohol described in part (**c**) would a person be safe to drive? (2)

f.   Name the disease of the liver caused by excessive consumption of alcohol. (1)

3. Each year in Britain as many as 750 people are paralysed as a result of damage to the spinal cord.

a. The pie chart shows the different causes of this damage.

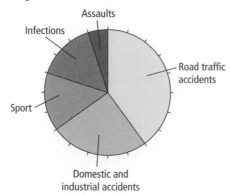

i. What is the most likely cause of damage? (1)
ii. What proportion of injuries to the spinal cord are caused by domestic and industrial accidents? Show your working. (2)
iii. Why is it important that young people are vaccinated against meningitis? (1)
b. Explain how it is possible that a girl with a damaged spinal cord cannot feel the shoes she is wearing but is able to write quite easily. (2)

4. The diagram shows a section of the spinal cord, illustrating the pathway of a simple reflex arc.

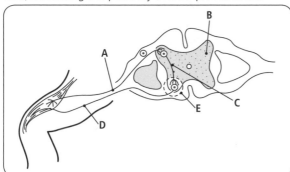

a. Identify the structures labelled **A** to **E**. (5)
b. Draw a simple diagram to explain how impulses cross the gap labelled **E**. (3)
c. Some drugs have their effects by affecting natural chemicals found in this gap. Several of these drugs are described in the table below.

	DRUG	EFFECT ON GAP
A	Caffeine	Prevents breakdown of natural chemical after it has crossed gap
B	Marijuana	Blocks natural chemical before it can cross gap
C	Heroin	Mimics (copies) natural chemical
D	Cocaine	Prevents removal of natural chemical after it has crossed the gap
E	Nicotine	Mimics (copies) natural chemical

State whether each of these drugs will act as a stimulant or as a sedative. Explain your answer. (5, 5)

5. A group of students suggested that coffee is enjoyable because it speeds up the heart rate. They gave several groups of people different amounts of coffee one morning and collected the following information.

NUMBER OF CUPS OF COFFEE DRUNK	HEART RATE (BEATS PER MINUTE) TOTAL	MEAN
0	74, 76, 72, 72, 78, 68	
1	78, 78, 82, 72, 72, 70	
2	78, 78, 79, 87, 80, 72	
3	80, 82, 78, 81, 78, 76	
4	76, 78, 88, 90, 88, 86, 78	
5	80, 90, 88, 88, 94, 92	

a. i. Copy and complete the table by calculating the mean heart rate for each of the test groups.
ii. Present your data in a suitable graphical form.
iii. Does this data support the hypothesis that coffee affects the heart rate?
iv. Suggest three precautions that the students should have taken to ensure that their data were valid.
b. How could you use epidemiology to investigate the hypothesis that heroin dependence in women leads to lower birth mass of children? Why is an epidemiological approach necessary to use in this sort of investigation?

6. Study the diagram below, then answer the questions that follow.

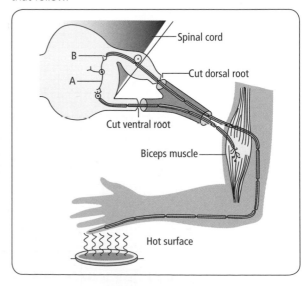

a. Name the type of neurone labelled **A**. What important function does it have?
b. Name the gap labelled **B**.

Supplement

c. In what form is a message transmitted:
    i.   along a nerve fibre
    ii. across the gap **B**?
d. Give **two** examples of spinal reflexes, **two** of cranial reflexes and **one** conditioned reflex.
e. How would sensation in the limb be affected if:
    i.   the dorsal root (branch) was cut
    ii. the ventral root was cut?
f. Would a reflex action occur in the limb when the dorsal root was being cut? Explain your answer.

g. Name the parts **C** to **G** on the diagram of a sensory neurone below. State two ways in which this neurone differs from a motor neurone.

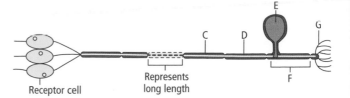

Receptor cell   Represents long length   F

REVISION SUMMARY:
Match the terms with the correct definitions

	TERM		DEFINITION
A	Neurone	1	Junction between nerve cells
B	Stimulus	2	An electrical insulator for the neurone
C	Synapse	3	A structure which can detect a stimulus
D	Neurotransmitter	4	An automatic action with a positive survival value
E	Myelin	5	A conducting cell
F	Receptor	6	A structure which can carry out an action
G	Effector	7	Region of brain which matches stimuli to responses
H	Reflex	8	The brain and spinal cord
I	Association centre	9	A change in the environment
J	CNS	10	Chemical which bridges the synapse

The eye is a sense organ

Different tissues work together to perform one function.

- **Retina** contains light sensitive cells.
- **Cornea** and **lens** focus the light onto the retina.
- **Tough outer coat** and **iris** protect and support the eye.

SENSE ORGANS ARE GROUPS OF RECEPTOR CELLS		
ORGAN	RESPONDS TO	SENSE
Eye	light	sight
Ear	sound	hearing
Nose/tongue	chemicals	smell/taste
Skin	touch	touch/feeling
Skin/part of brain	temperature	hot and cold

Accommodation: is the process of producing a finely focused image on the retina. It is carried out by the action of the **ciliary muscles** on the **lens**.

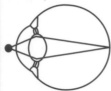

Close object:
Light must be **greatly refracted** (bent), ciliary muscles **contract**, pull eyeball inwards, ligaments **slack**, lens becomes **short and fat**.

Distant object:
Light can be **less refracted**, ciliary muscles **relax**, eyeball becomes spherical, ligaments **tight**, lens is pulled **long and thin**.

Rods: provide **black-and-white** images but, because several may be 'wired' to a single sensory neurone in the optic nerve, they provide **great sensitivity** at low light intensity (night vision) but images lack detail.

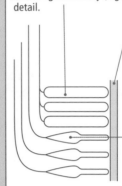

Layer of pigment: prevents internal reflection which might lead to 'multiple/blurred' images.

Cones: provide very **detailed** images, in **colour** (there are three types, sensitive to red, green and blue light) but **only under high light intensity** (their connections to the optic nerve make them rather insensitive).

Lens	Ciliary muscle	Suspensory ligament

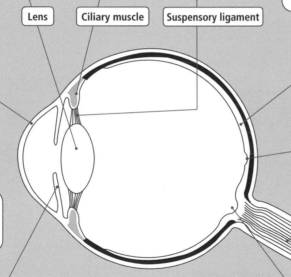

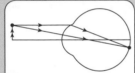

Cornea: a transparent layer that is responsible for most of the refraction (bending) of light rays from the object to form the inverted, and smaller, image on the retina.

Retina: contains the light-sensitive cells, the rods and cones. Light arriving at this layer will produce an **inverted, smaller image**.

Yellow spot (fovea): has the greatest density of cones and thus offers **maximum sharpness** but only works at full efficiency in **bright light**.

Pupil: the circular opening that lets light into the eye. Appears 'black' because the choroid is visible through it.

Optic nerve: composed of sensory neurones that carry nervous impulses to the **visual centre** at the rear of the brain.

Iris: the 'coloured' part of the eye, which may expand and contract to control the amount of light which enters the eye – this is a **reflex action**.

Blind spot: corresponds to the exit point for the optic nerve. There are no light-sensitive cells here so that light falling on this region cannot be detected.

The pupil reflex

Low light intensity: radial muscles of iris contract and the pupil is opened wider – **more light may enter and reach retina**.

High light intensity: circular muscles of iris contract and the pupil is reduced in size – **less light may enter** and **retina is protected from bleaching**.

The radical and circular muscles of the iris are the antagonistic muscles – they have opposing actions.

1. Jasmine went into a dark room from a bright corridor.

   a. Jasmine's right eye before and after entering the dark room is shown.

   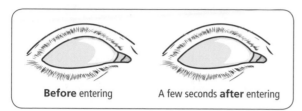

   Before entering   A few seconds **after** entering

   i. Copy and complete the diagrams by **drawing** the appearance of the pupil and iris

   1 before entering the dark room, (1)

   2 a few seconds after entering the dark room. (1)

   ii. Label the following parts of the eye on the first diagram.

   iris   pupil   sclera (3)

   b. Explain how the size of the pupil was changed when Jasmine went into the dark room. (2)

   c. Explain why Jasmine could see shapes but **not** colours in the dark room. (3)

   Cambridge IGCSE Biology 0610 Paper 3 Q4 November 2005

2. The diagram shows a section through the human eye.

   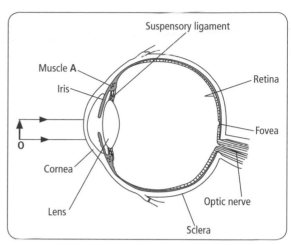

   a. Which **two** parts of the eye help to bend the light rays to bring them to a focus? (1)

   b. If object **O** were moved closer to the eye, what would muscle **A** do and how would this help to bring the light to a focus? (2)

3. Match the terms with their definitions.

	TERM		DEFINITION
A	Retina	1	Receptors that can distinguish red from green from blue
B	Rods	2	Structure that adjusts, focusing light onto the retina
C	Cones	3	Muscular structure that adjusts the amount of light entering the eye
D	Cornea	4	The opening that allows light to reach the retina
E	Lens	5	The light-sensitive layer in the eye
F	Optic nerve	6	Muscular structure that allows accommodation
G	Pupil	7	Exit point for the optic nerve
H	Iris	8	Carries sensory information to the visual association centre
I	Ciliary body	9	Receptors that are sensitive to low levels of light
J	Blind spot	10	Transparent layer that refracts light

(10)

4. Below is a section through the eye with a ray of light passing through it and four muscles labelled **A**, **B**, **C** and **D**.

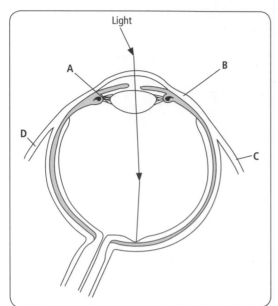

a. Complete the table.

PART	NAME OF MUSCLE	EFFECT OF CONTRACTION
A		allows the lens to become fatter for focusing on close objects
B	iris circular muscle	

(2)

Muscles **C** and **D** are voluntary muscles that are antagonistic. They are attached to the eye socket of the skull.

b. i. Explain the terms *voluntary* and *antagonistic*. (2)
   ii. Suggest the effect on the eye when muscle **C** contracts. (1)
   iii. Explain how the eye would return to its original position after this contraction. (2)

c. Light passes through parts of the eye to reach the retina. Complete the flow chart by putting the following terms in the boxes to show the correct order that the light passes through them.

aqueous humour   **cornea**   **lens**   **pupil**   **vitreous humour**

☐ → ☐ → ☐ → ☐ → ☐ → retina

(2)

d. The retina contains rods and cones. Complete the table to distinguish between rods and cones.

	TYPE OF LIGHT DETECTED	DISTRIBUTION IN THE RETINA
Rods		
Cones		

(4)

Cambridge IGCSE Biology 0610 Paper 3 Q2 June 2005

5. Below is a section through the retina of a human eye.

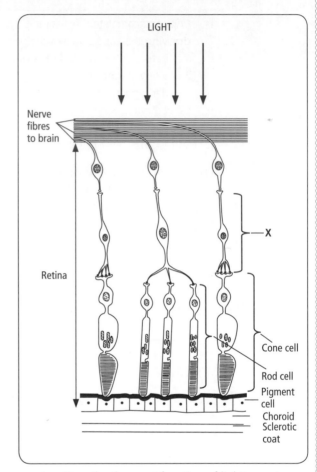

a. i. Describe the exact function of light receptors in terms of conversions they bring about. (2)
   ii. Describe how the functions of the **rods** are different from those of the **cones**. (2)

b. What **type** of cell is **X** and how is its structure related to its function? (4)

c. Through what structures will light rays pass from the outside of the eye until they stimulate the rods/cones. (3)

d. A bright light is suddenly shone into a person's eyes. The person raises his forearm to shield his eyes from the glare.
   i. Describe how antagonistic muscles in the arm will bring about this response. (5)
   ii. Describe how the eye is linked with the effector to produce this response. (5)

RECEPTORS AND THE SENSES: Crossword

ACROSS:
2 Light-sensitive layer of the eye
5 Sense that depends on the semi-circular canals
6 The organ responsible for 9 across
7 A reflex action that might be set off by touch receptors in the trachea
9 The sense of vision
11 Photoreceptors responsible for colour vision
12 Tighten during viewing of a distant object
15 Automatic action with a positive survival value
18 Photoreceptors that are most common around the edge of the retina
19 Location of chemoreceptors responsible for the sense of smell
22 A sense with receptors located in the skin
25 With 1 down – a structure responsible for detecting changes in temperature
26 Receptor that starts the knee jerk reflex
28 Effector responsible for controlling the shape of the lens (7, 6)
31 A change in the environment that can be detected by a receptor
32 Location of receptors concerned with touch and heat
33 Muscle that controls the amount of light entering the eye
34 Does this layer give you a black eye?

DOWN:
1 Structure capable of detecting a stimulus
3 Location of the chemoreceptors responsible for taste
4 Crystalline structure that 'fine focuses' light onto the retina
8 Is this spot yellow?
10 Sense related to the ears
13 The ability to link a sensory input with the correct motor output
14 There are four separate types of this sense, all started by receptors on the tongue
16 A chemical sense that helps to protect the breathing passages from dangerous chemicals
17 Sensory nerve from the ear
20 Sensory nerve from the eye
21 Has outer, middle and inner sections – all concerned with responding to sound stimuli
23 Transparent, refractive layer at the front of the eye
24 No photoreceptors here!
26 This type of neurone leaves a sense organ
27 Protective secretion for the eyes – especially when peeling onions
29 Sensation caused by overstimulation of any receptor
30 Circular opening to admit light to the rear of the eye

Chapter 22:
Hormones, the endocrine system and homeostasis

The endocrine system

The endocrine system uses chemical 'messengers' called **hormones** to co-ordinate the response of organs to stimuli and changes in the environment. A **hormone** is a chemical substance, released from a gland and carried by the blood which alters the activity of one of more specific target organs.

The **endocrine system** is made up of **glands** that can release special chemicals called hormones into the blood stream. Detectors on body organs, e.g. the heart, detect changes in the level of hormones in the blood. When the hormone level changes, the organ responds.

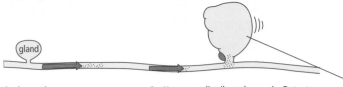

| 1 Endocrine gland responds to change in environment | 2 Gland releases hormone | 3 Hormone distributed throughout the body | 4 Detector on organ senses change in hormone level | 5 Organ responds, e.g. heart beats faster |

Glands, hormones and their functions

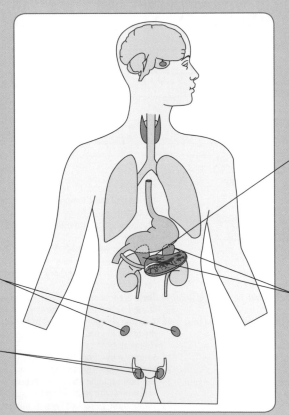

Pancreas: produces insulin, which controls the level of sugar in the blood.

Diabetes is a disease where the body cannot control the blood sugar level properly. **Diabetes type II** sufferers have to control their diet carefully. Sufferers from **type I diabetes** have to inject themselves with insulin.

Ovaries: in females the ovaries produce **oestrogen** and **progesterone**. These cause changes at puberty and control the menstrual cycle.

Testes: in males, the testes produce **testosterone** which causes changes at puberty.

Adrenal glands: produce adrenaline when there is danger. This prepares the body for fighting or running away (**fight** or **flight** response).
Adrenaline has many effects, for example:
• increases blood glucose concentration
• increases heart rate
• increases breathing rate

Comparing the nervous system with the endocrine system

	WHAT DETECTS CHANGE? (RECEPTOR)	WHAT IS THE 'MESSENGER'?	HOW IS THE MESSAGE CARRIED?	WHAT RESPONDS? (EFFECTOR)
NERVOUS SYSTEM	sensory receptors	electrical impulse	nerves – very fast	a muscle or a cell
ENDOCRINE SYSTEM	gland (via brain and bloodstream)	chemical (hormone)	bloodstream – slower	one or more organs

Supplement

Blood sugar regulation is an example of **homeostasis**. **Homeostasis** is the maintenance of a constant internal environment.

WHAT IS BLOOD SUGAR LEVEL?
Glucose is the cells main source of energy, and it must always be available to them...

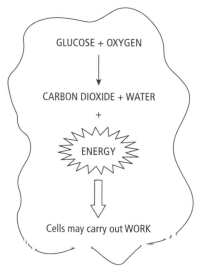

GLUCOSE + OXYGEN

↓

CARBON DIOXIDE + WATER

+

ENERGY

⇩

Cells may carry out WORK

... so that the body keeps a constant amount of glucose in the blood. This is the **blood sugar level** and is usually maintained at about 1 mg of glucose per cm^3 of blood.

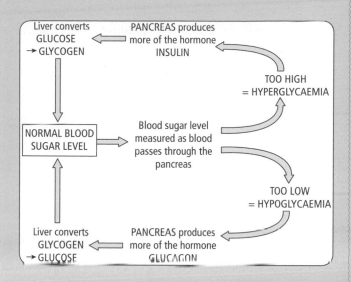

Liver converts GLUCOSE → GLYCOGEN

PANCREAS produces more of the hormone INSULIN

TOO HIGH = HYPERGLYCAEMIA

NORMAL BLOOD SUGAR LEVEL

Blood sugar level measured as blood passes through the pancreas

TOO LOW = HYPOGLYCAEMIA

Liver converts GLYCOGEN → GLUCOSE

PANCREAS produces more of the hormone GLUCAGON

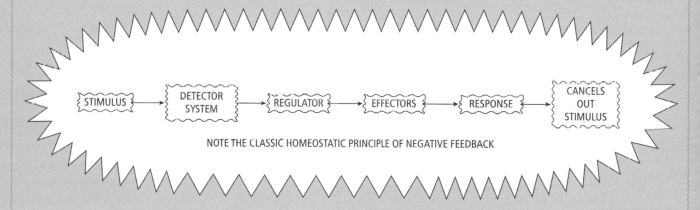

STIMULUS → DETECTOR SYSTEM → REGULATOR → EFFECTORS → RESPONSE → CANCELS OUT STIMULUS

NOTE THE CLASSIC HOMEOSTATIC PRINCIPLE OF NEGATIVE FEEDBACK

WHAT IS DIABETES?
Diabetes is a condition in which there are higher than normal blood glucose concentrations.

Type I diabetes is the result of the pancreas failing to secrete enough insulin.
- symptoms include:
  excessive thirst, hunger or urine production;
  sweet-smelling breath;
  high 'overflow' of glucose into urine (test with clinislix).
- long term effects if untreated include:
  premature ageing;
  cataract formation;
  hardening of arteries;
  heart disease.
- treatment is by regular injection of pure insulin – much of this is now manufactured by **genetic engineering**.

Type II diabetes can be controlled by adjusting diet to reduce sugar intake and does not require insulin injections.

1. Complete the table with the names of the three missing hormones. Choose your answers from the following list.

adrenaline   testosterone

auxin   oestrogen

follicle-stimulating hormone   progesterone

insulin

Hormone	What it does in the human body
	lowers the amount of sugar in the blood
	causes development of the female secondary sexual characteristics
	causes development of the milk-secreting glands in the breasts during pregnancy

(3)

2. Bovine somatotrophin (BST) is a natural growth hormone. The hormone can also be produced by genetic engineering, and is valuable because it can affect meat and milk production in cattle. The hormone can be injected into the bloodstream of cattle, or supplied in a gelatin coat so that it can be given in cattle food.

The bar chart shows the results of some experiments carried out at a cattle research institute.

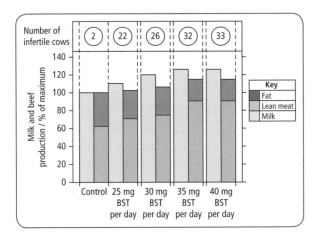

a. What are the advantages to a farmer of treating cattle with BST? (1)

b. Suggest a possible disadvantage to the farmer of treating cattle with BST. (1)

c. Suggest a possible disadvantage to the human consumers of beef treated with BST. (2)

d. Explain why enclosing the BST in a gelatin coat is necessary if the hormone is to be given in the cattle food.

e. i. Chickens are also treated with growth hormones. One such hormone is the female sex hormone, oestrogen. This hormone makes chickens grow more quickly and retain more water in their tissues, so they become heavier in a shorter time.

ii. Male body builders require a high protein diet. Many find that beef is too expensive for regular consumption, and so eat chicken and chicken products as often as twice a day. Body builders often report 'feminization', i.e. the growth of breasts, the shrinkage of their testes and the loss of facial hair. Explain how these two observations could be linked. (2)

3. One important hormone in mammals is **adrenaline**. This has many effects on body systems.

a. Match the columns below to show the benefits of adrenaline secretion.

	EFFECT OF ADRENALINE		BENEFIT TO BODY
A	Pupil of eye widens	1	Blood moves more quickly to working tissues
B	Heart beats faster and deeper	2	Excess heat can be lost
C	Glycogen is hydrolyzed	3	More oxygen enters the blood
D	More blood delivered to brain	4	More oxygen and glucose are delivered to respiring muscle cells
E	More blood delivered to skin	5	Allows more light onto the retina
F	Less blood delivered to the gut	6	More information can be processed and actions taken
G	Breathing rate increase	7	Digestion will be slowed
H	Blood flow to muscle is increased	8	More glucose is available for respiration

(7)

b. Use information from this table to explain why adrenaline is sometimes called the 'flight, fright or fight' hormone. (3)

4. The normal blood glucose level is 1 mg per cm³. Ten people with normal blood glucose levels were tested for blood glucose and plasma insulin levels over a period of six hours. The mean values for these measurements were calculated and recorded. The test period included two meals and a session of exercise. The results are shown in the table below.

TIME / HOURS	ACTIVITY	BLOOD GLUCOSE LEVEL / mg cm⁻³	PLASMA INSULIN LEVEL / μg cm⁻³
0	Meal eaten	1.0	10
0.5		1.5	20
1.0		1.0	40
1.5		0.8	25
2.0	Exercise started	0.8	15
2.5	Exercise finished	1.2	10
3.0		1.0	20
3.5		1.0	10
4.0	Meal eaten	1.0	10
4.5		1.4	20
5.0		1.0	35
5.5		0.8	40
6.0		0.8	10

a. Present all the data in the form of a graph.   (6)
b. What effect does the period of exercise have on the blood glucose and plasma insulin levels? Explain your answer.   (3)
c. Suggest two other hormones that would change in concentration in the blood during the period of exercise. Why are these hormones important?   (4)
d. Why is it good experimental technique to
    i.   take **mean** values for blood glucose and plasma insulin levels
    ii. use only subjects with normal glucose levels?   (2)
e. How long after a meal does it take for the blood glucose level to return to normal?   (1)

5. Insulin is a hormone produced to control blood glucose levels. Diabetics do not have a natural ability to control these levels.
a. Define the term *hormone*.   (2)
b. With reference to the pancreas and the liver, describe the role of insulin in controlling blood glucose levels.   (4)
c. •   Insulin is a protein.
    •   Diabetics can control their blood glucose levels artificially by injecting insulin.
    •   Many medicines are swallowed as tablets.
    Explain what would happen to the insulin in the stomach if it was swallowed as a tablet.   (2)

Cambridge IGCSE Biology 0610 Paper 3 Q4 June 2006

6. Read the following passage which is from an advice book for diabetics.

Hypoglycaemia or 'hypo' for short, occurs when there is too little sugar in the blood. It is important always to carry some form of sugar with you and take it immediately you feel a 'hypo' start. A hypo may start because:
•   you have taken too much insulin, or
•   you are late for a meal, have missed a meal altogether, have eaten too little at a meal, or
•   you have taken a lot more exercise than usual.
The remedy is to take some sugar.
An insulin reaction usually happens quickly and the symptoms vary – sweating, trembling, tingling of the lips, palpitations, hunger, pallor, blurring of the vision, slurring of speech, irritability, difficulty in concentration. Do not wait to see if it will pass off, as an untreated 'hypo' could lead to unconsciousness.

a. Many diabetics need to take insulin.
    i.   Explain why.   (2)
    ii. Explain why there is too little sugar in the blood if too much insulin is taken.   (3)
    iii. Explain why there is too little sugar in the blood if the person exercises more than usual.   (3)
b. Suggest why sugar is recommended for a 'hypo', rather than a starchy food.   (3)
c. Explain how the body of a healthy person restores blood sugar level if the level drops too low.   (3)
d. Explain, using insulin as an example, what is meant by negative feedback.   (3)

REVISION SUMMARY: Fill in the gaps

Complete the paragraph by filling in the gaps with words from the list below. You may use each word once, more than once or not at all.

ADRENALINE, HORMONE, BLOODSTREAM, DIGESTIVE SYSTEM, ENDOCRINE ORGAN, INSULIN, TARGET ORGAN, TRACHEA, OESOPHAGUS

In all mammals there is a chemical coordination system. This system uses chemical messengers called ...., which are secreted by the ...., travel in the .... and have an effect on a .... An example of this type of chemical messenger is ...., which, during periods of fear or anxiety, relaxes the ...., allowing the person to breathe more easily.   (6)

Hormones and plant growth

Groups of plant hormones called **auxins** control the growth of shoots and roots.

PHOTOTROPISM

Plant shoots grow towards light. This is called phototropism. Phototropism is caused by auxins produced in the shoot tip. The auxins promote elongation of cells and are redistributed under the influence of light.

HOW IT WORKS

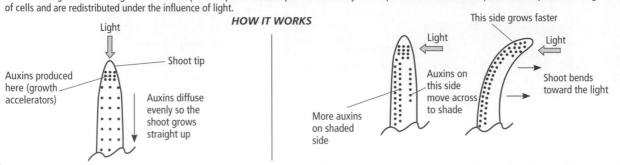

Light

Shoot tip

Auxins produced here (growth accelerators)

Auxins diffuse evenly so the shoot grows straight up

This side grows faster

Light

Light

Auxins on this side move across to shade

More auxins on shaded side

Shoot bends toward the light

GRAVITROPISM

Plant roots grow downwards in the direction of the pull of gravity. This is called gravitropism. Gravitropism is caused by auxins produced in the root tip. Root auxins slow down growth (i.e. opposite to the effect of shoot auxins).

HOW IT WORKS

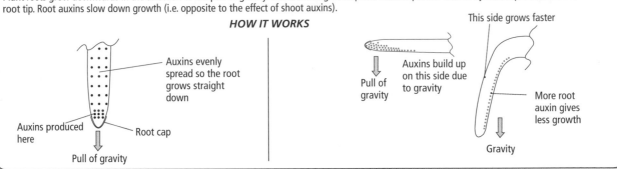

Auxins evenly spread so the root grows straight down

Auxins produced here

Root cap

Pull of gravity

This side grows faster

Pull of gravity

Auxins build up on this side due to gravity

More root auxin gives less growth

Gravity

USING PLANT HORMONES

Hormone rooting powders can promote root growth in plant cuttings.

Plant hormones can 'trick' flowers into forming fruits without fertilisation. This produces 'seedless' fruits.

Selective weed killers, such as the synthetic plant hormone 2, 4-D, are absorbed by broad-leaved weeds but narrow-leaved grasses and cereal crops absorb less. Hormones cause the weeds to grow so quickly that they wither and die, and the crop of cereals can grow better without competition.

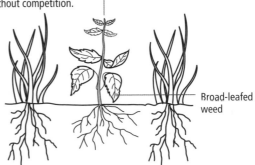

Broad-leafed weed

1. All living organisms respond to changes in their environment. The responses they show have an important effect on their survival. Plants respond by altering their direction of growth. Because growth is involved, plants respond more slowly than animals do. Complete the following sentences about plant responses.

    a. A tropism is defined as a change in the direction of .... of a plant in response to a directional ....

    b. In gravitropism, the plant is responding to ....

    c. The shoot of a plant will grow towards light, that is it shows .... phototropism. This response allows the plant to produce more food by .... (5)

2. Growing oat seedlings are placed in a box that only allows light from one side to reach them, as shown in the diagram below.

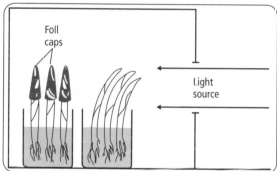

Which of the following conclusions is the correct one? Explain your answer. (2)

    The shoot with the foil cap:

    a. cannot photosynthesize without light, and so cannot grow in a curved shape.

    b. is too far away from the light to be stimulated.

    c. cannot respond because the tips must receive the stimulus.

    d. is deprived of carbon dioxide, and so cannot grow.

Supplement

3. Young oat seedlings were treated with a substance **X**. The substance **X** was given to the seedling by mixing it with lanolin paste and smearing the paste on one side of the seedling, as shown in the diagram below.

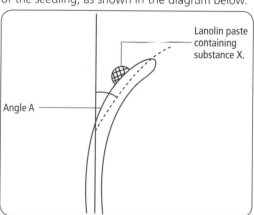

An experiment was carried out in which different concentrations of substance **X** were given to different seedlings, and the angle **A** measured in each case. The results are shown below.

CONCENTRATION OF X / mg dm⁻³	ANGLE A / DEGREES
1	3
2	6
4	10
7	16
10	19

    a. Plot these results in the form of a graph. (5)

    b. What would be the value of angle **A** at a concentration of **X** of 5 mg dm⁻³? (2)

    c. What is the name given to the response shown by the oat seedlings? (1)

    d. What is the name of substance **X**? (1)

4. Auxin is a hormone made by the tips of plant shoots. The diagram below shows the movement of auxin in two young shoots, **A** and **B**, which were treated in different ways. **X** shows where auxin was made. Both shoots were kept in the dark.

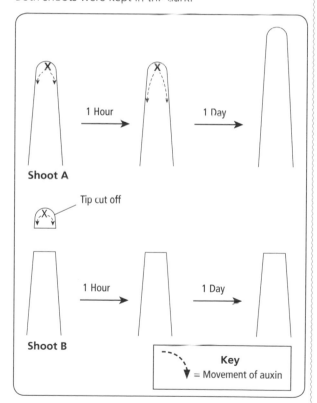

    a. Explain the difference in the growth of shoot **A** and shoot **B** at the end of one day. (4)

    b. A third shoot, **C**, was grown in a box so that light shone onto it from only one side. The diagram below shows movement of auxin in this shoot and the result of the experiment. (1)

Supplement

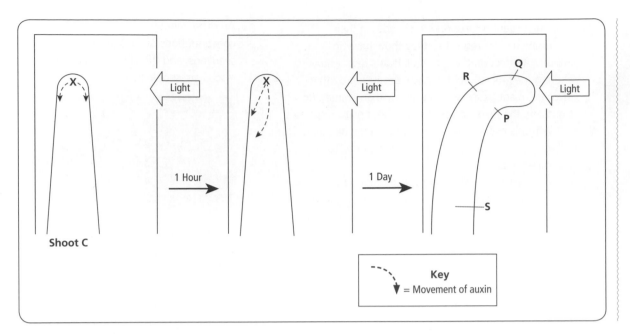

Shoot C

Key

= Movement of auxin

i. Describe the movement of auxin in shoot **C** after one hour.

ii. Auxin causes plant cells to elongate (grow longer). At which point, **P**, **Q**, **R** or **S**, would cells have elongated the most? (1)

c. Plant hormones are sometimes used by humans to control plant growth. Give **two** examples of this. (2)

5. a. State **three** conditions necessary for the germination of most seeds. (3)

A student carried out an experiment on the direction of growth of the root of a germinating seed and the shoot of a seedling. Figure A shows the experiment when first set up. The electric motors slowly turn the cork base and the plant pot Figure B shows the experiment after two days. The root and shoot received the same amount of light from all directions.

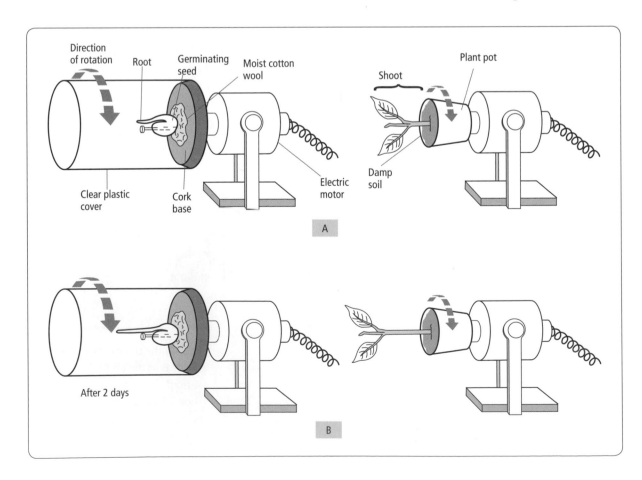

Direction of rotation   Root   Germinating seed   Moist cotton wool

Plant pot

Shoot

Electric motor

Damp soil

Clear plastic cover   Cork base

A

After 2 days

B

b. Suggest why a clear plastic cover was provided for the root but not for the shoot. (1)

c. i. Complete the diagram below to show how the root and shoot would appear 24 hours **after the motors were switched off**. (2)

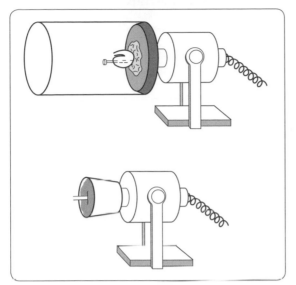

ii. Name the responses shown by the root and the shoot that you have drawn.

Root response: ....

Shoot response: .... (2)

iii. Suggest why these responses were **not** shown in Figure A. (1)

REVISION SUMMARY: Match the terms with the correct definition

	TERM		DEFINITION
A	Tropism	1	Form of chemical pest control, allowing crop plants to grow without competition
B	Auxin	2	Fusion of male and female gametes
C	Gravity	3	Plant response to light
D	Gravitropism	4	Hormone that increases the rate of root growth from the base of a cut stem
E	Phototropism	5	Directional growth response in a plant
F	Apex	6	Downward force resulting from the mass of the Earth
G	Rooting powder	7	Hormone affecting growth of both root and shoot
H	Weedkiller	8	Development of a seed into a young plant
I	Germination	9	Plant response to gravity
J	Fertilisation	10	The tip of a root or shoot

Chapter 24:
Reproduction

Reproduction is the generation of new individuals of the same species.

Reproduction can be either ASEXUAL or SEXUAL

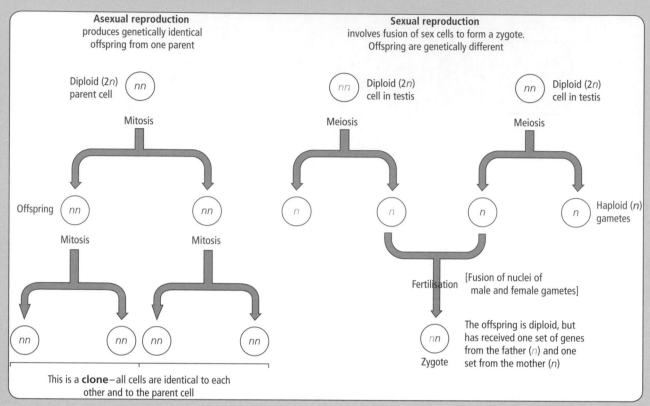

Asexual reproduction produces genetically identical offspring from one parent

Diploid (2n) parent cell — nn

Mitosis

Offspring — nn, nn

Mitosis, Mitosis

nn, nn, nn, nn

This is a **clone** – all cells are identical to each other and to the parent cell

Sexual reproduction involves fusion of sex cells to form a zygote. Offspring are genetically different

nn — Diploid (2n) cell in testis

Meiosis

n, n — gametes

nn — Diploid (2n) cell in testis

Meiosis

n, n — Haploid (n) gametes

Fertilisation [Fusion of nuclei of male and female gametes]

nn — Zygote

The offspring is diploid, but has received one set of genes from the father (n) and one set from the mother (n)

Sexual and asexual reproduction compared

Sexual reproduction and vegetative propagation both have advantages and disadvantages. Many plants make the best of both worlds, and reproduce both sexually and asexually.

	ADVANTAGE	DISADVANTAGE
Asexual	• Only one parent needed • Rapid colonisation of favourable environments	• No variation, so any change in environment conditions will affect all individuals
Sexual	• Variation, so new features of organisms may allow adaptation to new environments	• Two parents needed • Fertilisation is random, so harmful variations can occur

Which is better for growing crops?

SEXUAL:

• new varieties can be developed

• might give bigger yield, or tolerate difficult conditions

ASEXUAL:

• Varieties with useful features can be cloned

Flower structure is adapted for pollination

CARPEL: the female part of the flower.

Stigma: surface on which pollen grains, containing male gametes, may be deposited.

Style: stalk that holds stigma in prominent position, and down which pollen tube may grow.

Ovary: contains the ovule, which encloses the female gamete. Ovary wall may become part of the fruit.

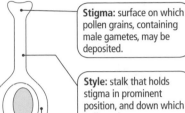

STAMEN: the male part of the flower.

Anther: produces pollen grains, containing male gametes, within the pollen sacs.

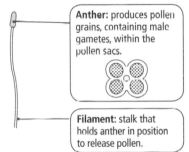

Filament: stalk that holds anther in position to release pollen.

PETALS: play an important part in pollination.

SEPALS: protect the other floral parts against drying out and fungal attack. There are the same number as there are petals, and they are usually green in colour.

POLLINATION is the transfer of pollen from the stamen to the stigma.

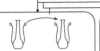

Cross-pollination involves pollen transfer between different flowers of the same species. It involves genetic variation but may be risky because it requires a **vector** (a carrier).

Self-pollination involves pollen transfer from stamen to stigma of the same flower. It is less risky but it limits the chances of genetic variation.

Wind-pollinated species usually occur in dense groups, e.g. the grasses.		Insect-pollinated species are usually solitary or in small groups.
Dull in colour. Small, or even absent to reduce obstruction of pollen access to stigma.	PETALS	Large, brightly coloured, may be scented and/or have guidelines. Base may produce attractive **nectar.**
Long and flexible so that pollen may be easily released.	STAMENS	Short and stiff to brush pollen against body of visiting insect.
Long and feathery, giving a large surface area to receive pollen.	STIGMA	Held inside petals to ensure contact with body of visiting insect.
Small, dry, enormous quantities.	POLLEN	Large, sticky, small amounts.

RECEPTACLE: the swollen tip of the flower stalk. It is the base on which the other parts of the flower stand.

1. The diagram below shows a bee that collects food materials from some flowers belonging to the same species. While it does this the bee also assists in the reproductive processes of the flowers.

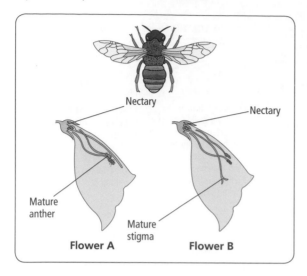

Flower A   Flower B

a. i.   Name the stage in the reproduction of the plants in which the bee is involved. (1)
    ii. Suggest how this process might take place between flowers **A** and **B**. (3)
b. The ovules in each flower can develop into seeds.
    i.   Which reproductive process must happen inside an ovule before it can become a seed? (1)
    ii. State which part of the flower develops into a fruit. (1)
c. Explain why plants grown from the seeds produced by these flowers will be similar to each other but may not be identical. (4)

Cambridge IGCSE Biology 0610 Paper 2 Q6 November 2005

2. A common garden weed, the Creeping Buttercup, can reproduce asexually by means of runners. The process is summarized in the diagram below.

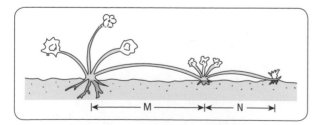

A horticultural student decided to investigate the growth of this weed. He took ten different buttercup plants and measured the lengths of the runners M and N. The results are shown in the table.

PLANT NUMBER	LENGTH OF RUNNER M / mm	LENGTH OF RUNNER N / mm
1	175	140
2	210	130
3	305	130
4	320	170
5	300	125
6	170	120
7	230	135
8	300	125
9	260	145
10	200	100

a. Calculate the mean lengths for runners **M** and **N**. Show your working. (4)
b. Calculate the difference in mean length between runners **M** and **N**. (2)
c. Suggest why there are differences between the length **M** in the ten different plants. (2)
d. The student was trying to develop a weedkiller that could control the buttercups. He noticed that one of the plants was very resistant to the weedkiller. Explain how this method of reproduction would allow the buttercup to survive throughout the garden, even if weedkiller is used. (2)

3. a. What is the difference between asexual and sexual reproduction in terms of the number of parents involved and the amount of variation in the offspring? (2)
b. What are the advantages and disadvantages of these differences? (4)
c. Some flowering plants use runners to reproduce asexually. Describe how they do this. (4)

4. The figure below shows a whole flower of *Nicotiana* (**A**) and a section of the same flower (**B**).

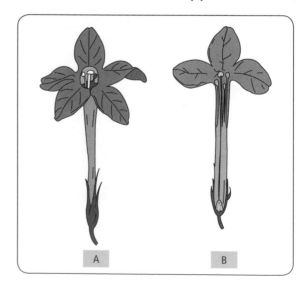

A   B

The figure below shows a whole flower of *Lilium* (**C**) and a section of the same flower (**D**).

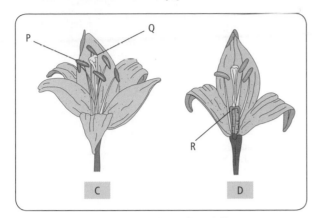

a. Complete the table to show **five** differences between the *Nicotiana* flower and the *Lilium* flower. On each line you should compare **one** feature of the flowers. The first one is done for you.

NICOTIANA FLOWER	LILIUM FLOWER
1 Fewer stamens are present	More stamens are present
2	
3	
4	
5	
6	

(5)

b. i. Compare the functions of structures labelled **P** and **Q**. (2)

ii. Explain why reproduction in flowers is considered to be **sexual** rather than **asexual** reproduction. (2)

c. State **four** reasons why it is very unlikely that these flowers are wind-pollinated. (4)

REPRODUCTION IN PLANTS: Crossword

ACROSS:
4 Landing platform for male sex cells
7 What the ovary develops into after fertilisation
9 An organism with both male and female sexual organs
10 Can be an agent of pollination or germination
12 Contains an embryo, a food store and a waterproof coat
13 Where the pollen is produced
15 Produces a nutritious, sugary secretion
17 Contains the female sex cells
18 The spreading out of offspring from a parent plant
19 The transfer of male sex cells to female flower parts
20 A flying pollinator
22 The main advantage of sexual reproduction
23 Stalk that supports 13 across in the male part of the flower
24 Underground stem for asexual reproduction – potato, for example

DOWN:
1 The female sex cell
2 Can be avoided if 18 across is efficient
3 A protective outer covering for a flower bud
5 Reproduction without fertilisation
6 The fusion of male and female sex cells
8 Athletic structure for asexual reproduction?
11 May be coloured to attract insects
14 The development of a seed into a young plant
16 Contains the male gamete
21 The pollen grain must send a tube down this structure

Chapter 26:
Germination and plant growth

Fertilisation, fruits and seed germination

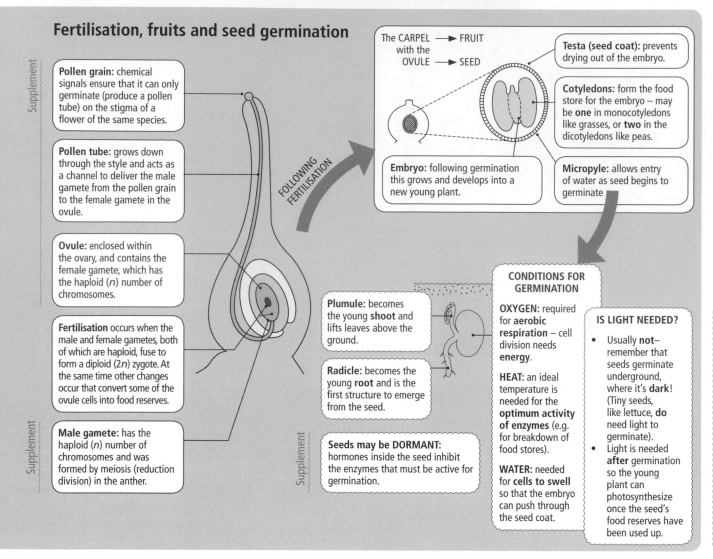

Pollen grain: chemical signals ensure that it can only germinate (produce a pollen tube) on the stigma of a flower of the same species.

Pollen tube: grows down through the style and acts as a channel to deliver the male gamete from the pollen grain to the female gamete in the ovule.

Ovule: enclosed within the ovary, and contains the female gamete, which has the haploid (*n*) number of chromosomes.

Fertilisation occurs when the male and female gametes, both of which are haploid, fuse to form a diploid (2*n*) zygote. At the same time other changes occur that convert some of the ovule cells into food reserves.

Male gamete: has the haploid (*n*) number of chromosomes and was formed by meiosis (reduction division) in the anther.

FOLLOWING FERTILISATION

The CARPEL → FRUIT with the OVULE → SEED

Testa (seed coat): prevents drying out of the embryo.

Cotyledons: form the food store for the embryo – may be **one** in monocotyledons like grasses, or **two** in the dicotyledons like peas.

Embryo: following germination this grows and develops into a new young plant.

Micropyle: allows entry of water as seed begins to germinate

Plumule: becomes the young **shoot** and lifts leaves above the ground.

Radicle: becomes the young **root** and is the first structure to emerge from the seed.

Seeds may be DORMANT: hormones inside the seed inhibit the enzymes that must be active for germination.

CONDITIONS FOR GERMINATION

OXYGEN: required for **aerobic respiration** – cell division needs **energy**.

HEAT: an ideal temperature is needed for the **optimum activity of enzymes** (e.g. for breakdown of food stores).

WATER: needed for **cells to swell** so that the embryo can push through the seed coat.

IS LIGHT NEEDED?

- Usually **not**– remember that seeds germinate underground, where it's **dark**! (Tiny seeds, like lettuce, **do** need light to germinate).
- Light is needed **after** germination so the young plant can photosynthesize once the seed's food reserves have been used up.

---

1. The diagram shows a section through half of a broad bean seed.

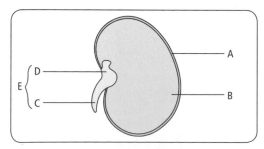

   a. Name all of the parts labelled **A** to **E** on the diagram. (5)
   b. Structure **B** contains a food store. A teacher explained that the food store provided energy for the growing seed. One student thought that the food would be fat, and another thought that it would be carbohydrate. Describe a test that you could carry out to decide who was right (or whether both were!).

Name the reagents that you would use, the steps you would take and the results you might expect. (5)

2. Seeds will only germinate if the environmental conditions are suitable. The experiment shown below was set up to investigate the conditions required for germination.

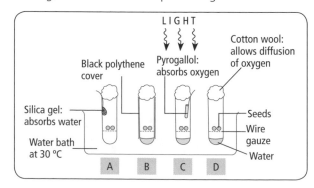

   a. In which of the tubes **A** to **D** will the seeds germinate? Explain your answers. (8)

b. The following results were obtained from an experiment to demonstrate that seed germination is affected by temperature. Seven lots of 100 seeds were kept on moist cotton wool and subjected to different temperatures. After 48 hours the number showing germination (at least 1 mm of root showing) were counted.

TEMPERATURE / °C	PERCENTAGE GERMINATION
5	3
15	22
25	55
35	79
45	52
55	19
65	2

i. Present the results in a graph. (5)
ii. What does the shape of the graph tell you about the control of germination in mustard seeds? (3)
iii. Why do gardeners sometimes scratch a hole in the seed coat before they plant the seeds? (2)

3. The diagram below shows a germinated seed.

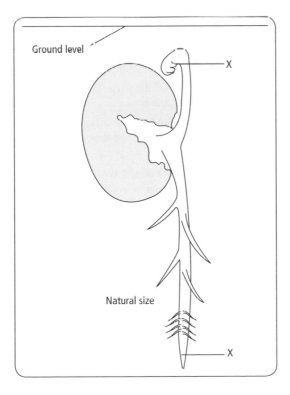

Ground level

X

Natural size

X

a. i. What food substance, stored in the seed, will be broken down into sugars? (1)
   ii. What food substance, stored in the seed, will be broken down into amino acids? (1)
b. Which class of chemicals will cause the stored food to break down into soluble compounds? (1)
c. Explain why amino acids are moved to the areas marked **X** in the diagram. (2)
d. Give **three** conditions that must have been present for this seed to germinate. (3)
e. i. Name fully the growth response being shown by the young shoot in the diagram. (2)
   ii. Explain the importance of this response to the plant. (1)

4. a. All seeds need oxygen, water and a suitable temperature to germinate. 22 °C is a suitable temperature for the germination of pea seeds. Light and dark conditions have no effect on pea seed germination. The diagram below shows an experiment on germination of pea seeds.

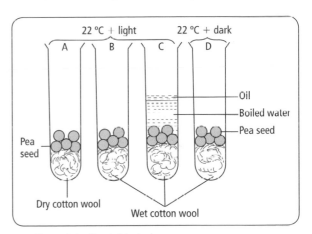

22 °C + light   22 °C + dark

A   B   C   D

Oil
Boiled water
Pea seed

Pea seed

Dry cotton wool   Wet cotton wool

Complete the table below.

TUBE	WOULD SEEDS GERMINATE? (WRITE YES OR NO)
A	
B	
C	
D	

(4)

b. The next table shows how the dry mass of barley seedlings changed over the first 35 days after sowing.

TIME AFTER SOWING / DAYS	0	7	14	21	28	35
DRY MASS / g	4.0	2.8	2.8	4.4	9.6	17.8

i. Complete the plotting of data from the table on the graph below. The first two results have already been plotted for you. (4)

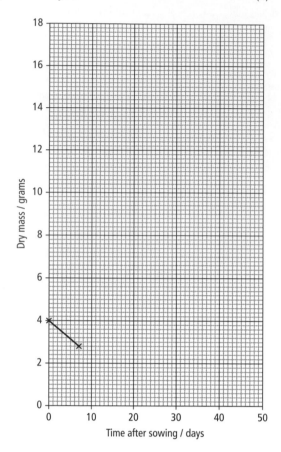

ii. How many days after sowing did the barley seedlings regain their original dry mass? (1)

iii. How many days after sowing did the barley seedlings **treble** their original dry mass? (1)

5. Two batches of tomato seeds were placed in dishes to germinate. Water was added to the seeds in dish **A**. A 25% solution of the juice from a fresh tomato was added to the seeds in dish **B**.

The diagram shows the two batches after 4 days.

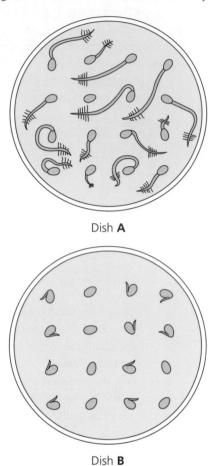

Dish **A**

Dish **B**

a. State **two** conditions, other than water, that are needed for the germination of the seeds. (1)

b. i. Describe briefly the results shown:
    in dish **A**
    in dish **B**

    ii. Make an enlarged, labelled drawing of one of the seedlings from dish **A**. (5)

c. i. Suggest what has caused the difference between the two batches of seeds.

    ii. Suggest how you would find out whether this factor, which affects the germination of tomato seeds, also affects the germination of other types of seeds. (4)

GERMINATION AND PLANT GROWTH: Crossword

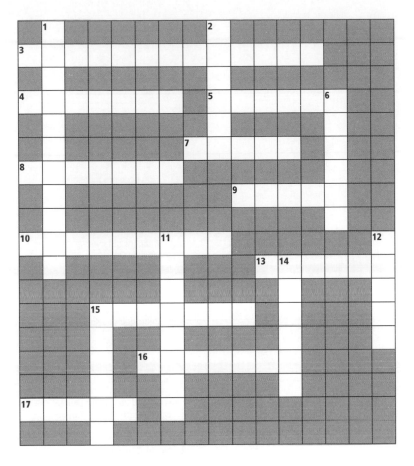

ACROSS:
3 The fusion of male and female gametes
4 Enzyme necessary to break down starch stores in a seed
5 The part of the seed that develops into a young plant
7 A coat that prevents a seed from drying out
8 Part of the embryo that develops into a root
9 Contains the female gamete and develops into the seed
10 A seed leaf, forming a food store in a seed
13 Specialized leaves for sexual reproduction in plants
15 Part of the embryo that becomes the shoot
16 Catalysts affected by temperature in germinating seeds
17 Needed for cells to swell so the embryo can push through the seed coat

DOWN:
1 The development of a seed into a young plant
2 A sex cell, with the haploid number of chromosomes
6 Gas required for aerobic respiration in germinating seeds
11 A condition in which a seed may fail to germinate
12 Develops from the ovary following fertilisation
14 An enzyme that breaks down food stores in fatty seeds
15 Grain that contains the male gamete

MALE REPRODUCTIVE SYSTEM

- produces **testosterone**
- manufactures **male gametes**
- transfers male gametes to female

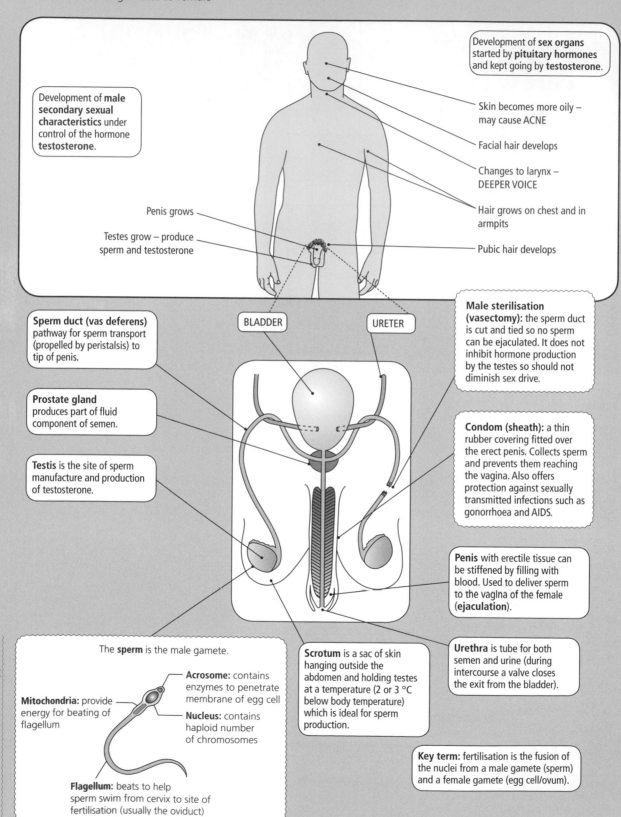

Development of **sex organs** started by **pituitary hormones** and kept going by **testosterone**.

Development of **male secondary sexual characteristics** under control of the hormone **testosterone**.

Skin becomes more oily – may cause ACNE

Facial hair develops

Changes to larynx – DEEPER VOICE

Hair grows on chest and in armpits

Pubic hair develops

Penis grows

Testes grow – produce sperm and testosterone

BLADDER

URETER

Sperm duct (vas deferens) pathway for sperm transport (propelled by peristalsis) to tip of penis.

Male sterilisation (vasectomy): the sperm duct is cut and tied so no sperm can be ejaculated. It does not inhibit hormone production by the testes so should not diminish sex drive.

Prostate gland produces part of fluid component of semen.

Condom (sheath): a thin rubber covering fitted over the erect penis. Collects sperm and prevents them reaching the vagina. Also offers protection against sexually transmitted infections such as gonorrhoea and AIDS.

Testis is the site of sperm manufacture and production of testosterone.

Penis with erectile tissue can be stiffened by filling with blood. Used to deliver sperm to the vagina of the female (**ejaculation**).

The **sperm** is the male gamete.

Scrotum is a sac of skin hanging outside the abdomen and holding testes at a temperature (2 or 3 °C below body temperature) which is ideal for sperm production.

Urethra is tube for both semen and urine (during intercourse a valve closes the exit from the bladder).

Acrosome: contains enzymes to penetrate membrane of egg cell

Mitochondria: provide energy for beating of flagellum

Nucleus: contains haploid number of chromosomes

Flagellum: beats to help sperm swim from cervix to site of fertilisation (usually the oviduct)

Key term: fertilisation is the fusion of the nuclei from a male gamete (sperm) and a female gamete (egg cell/ovum).

FEMALE REPRODUCTIVE SYSTEM

- produces **oestrogen / progesterone**
- develops **female gametes**
- accepts sperm
- allows development / birth of fetus

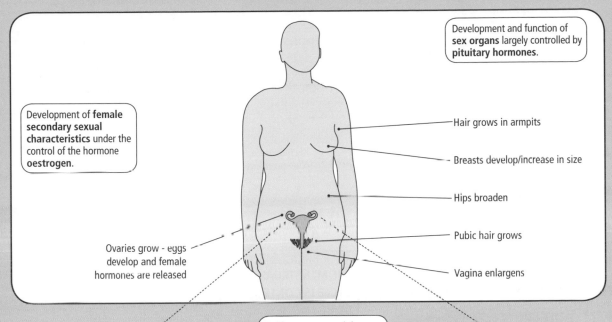

Development and function of **sex organs** largely controlled by **pituitary hormones**.

Development of **female secondary sexual characteristics** under the control of the hormone **oestrogen**.

Hair grows in armpits

Breasts develop/increase in size

Hips broaden

Pubic hair grows

Vagina enlargens

Ovaries grow - eggs develop and female hormones are released

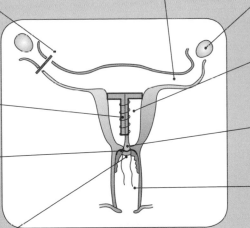

Oviduct: eggs travel down this tube to the uterus - fertilisation usually occurs in the first third of the oviduct.

Female sterilisation: the oviducts are cut or tied so that no eggs can be passed down to be fertilised.

Ovary: site of egg development and production of oestrogen and progesterone.

Coil (I.U.D. – Intra-Uterine Device): plastic with copper wire which irritates the lining of the uterus and prevents implantation of a fertilised egg.

Uterus: site of implantation of embryo and development of fetus.

Cervix: sperm deposited here during intercourse.

Cap (diaphragm): a thin rubber shield fitted over the cervix before intercourse. It prevents entry of sperm.

Vagina: accepts the penis during intercourse. Baby passes down the vagina during birth.

Spermicide (cream): is placed inside the vagina before intercourse – kills sperm. Usually used in conjunction with the diaphragm.

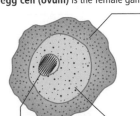

The **egg cell (ovum)** is the female gamete.

Jelly coat: changes at fertilisation so only one male nucleus can enter the egg cell

Cytoplasm: contains energy stores for the early development of the **zygote**/ball of cells

Nucleus: contains the **haploid** of chromosomes

Supplement

The menstrual cycle

This is a regular series of changes to the female reproductive system in preparation for fertilisation and pregnancy. It is controlled by **hormones** from the **pituitary gland** and the **ovary**.

In humans the cycles of the two ovaries are out of phase, so **each ovary** ovulates every 56 days but **each female** ovulates (releases an egg) every 28 days.

The uterus lining begins to degenerate **unless embryo implantation** has occurred when **progesterone** (from the corpus luteum) keeps the lining intact to begin pregnancy.

Uterus lining and its blood vessels are well-developed to receive an embryo. This optimum set of conditions remains for 6–7 days after ovulation.

A TYPICAL MENSTRUAL CYCLE

Stage 4 PRE-MENSTRUAL PHASE
Stage 1 MENSTRUATION
Stage 3 RECEPTIVE PHASE
Stage 2 REPAIR PHASE (of uterus lining)

Separation of uterus lining – blood and fragments of tissue leave the body via the vagina. This monthly 'period' occurs only in primates such as humans.

Ovulation, the release of the egg from the Graafian follicle into the oviduct. This is stimulated by **luteinizing hormone** and the release of fluid and an associated 'blip' in body temperature (see right) means that some human females are aware that ovulation has occurred.

Brain – receives information from other parts of the body, processes it and then 'instructs' the pituitary gland.

Pituitary gland – releases hormones which control activity of the ovary.

Luteinizing hormone – stimulates release of mature ovum from ovary.

Luteinizing hormone – stimulates development of corpus luteum from the remains of the follicle.

CONTRACEPTION AND THE MENSTRUAL CYCLE

- **The pill**: an oral dose of one or both hormones **oestrogen** and **progesterone** that acts as a **feedback inhibitor** of the release of **luteinizing** hormone by the pituitary gland. Thus **ovulation cannot occur** and **no pregnancy can result**.

- **The rhythm method** assumes that the time up to two days before ovulation and from four days after ovulation are **safe**, i.e. there should be no eggs available to be fertilised. This method is unreliable.

ARTIFICIAL INSEMINATION

- sperm is collected from donor
- can be frozen and stored
- thawed and then a syringe is used to place the sperm close to CERVIX

IVF: IN VITRO FERTILISATION

- sperm and egg allowed to fuse in test tube or petri dish
- zygote allowed to divide, then ball of cells allowed to implant in uterus
- zygote can be checked for genetic abnormality before implantation

THE MENSTRUAL CYCLE IS CONTROLLED BY HORMONES

FSH   LH   LH

Follicle stimulating hormone – stimulates development of Graafian follicle in the ovary.

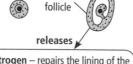

Graafian follicle

releases

Oestrogen – repairs the lining of the uterus and stimulates development of female sexual characteristics.

ovum

corpus luteum

releases

Progesterone – keeps the lining of the uterus ready for implantation and pregnancy.

Blood concentration of ovarian hormones

0   14   28

The lining of the uterus gradually thickens and develops more blood vessels ready to receive a fertilised egg.

Condition of lining of uterus

0   14   28

| MENSTRUATION | REPAIR PHASE | RECEPTIVE PHASE | PRE-MENSTRUAL PHASE |

'SAFE PERIOD'

Body temperature/°C

'SAFE PERIOD'

This 'blip' in temperature corresponds to ovulation.

0   14   28

ANTE-NATAL CARE

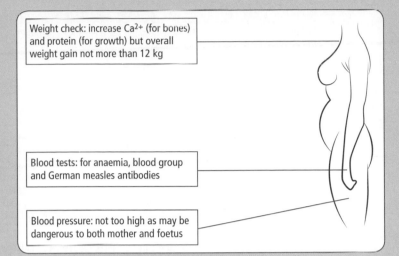

Weight check: increase Ca²⁺ (for bones) and protein (for growth) but overall weight gain not more than 12 kg

Blood tests: for anaemia, blood group and German measles antibodies

Blood pressure: not too high as may be dangerous to both mother and foetus

Labour and birth
At the end of pregnancy, contractions of the muscles in the uterus walls will cause the amniotic sac to break and the cervix to dilate. When the cervix is wide enough, the baby's head is pushed out through the vagina which acts as a birth canal. Once the baby starts breathing, the umbilical cord is tied (to prevent bleeding) and cut. The placenta then comes away from the wall of the uterus and leaves the vagina as the afterbirth.

Limit (or stop altogether) intake of alcohol, and other drugs, and stop smoking

BREAST IS BEST

MILK

MILKO MAX

Formula (artificial milk) is based on cow's milk
High fat
Too much protein
Very high sodium concentration
Very high lactose (milk sugar)

Natural (mother's) milk
Cheap
Contains antibodies
Ideal fat, protein, sodium and lactose content

and suckling may have a contraceptive effect!

Supplement

The placenta

This is responsible for protection and nourishment of the developing fetus.

PLACENTA: this disc-shaped organ has a number of functions:
- exchange of soluble materials between mother and fetus
- physical attachment of the fetus to the wall of the uterus
- protection
  1) of fetus from mother's immune system
  2) against dangerous fluctuations in mother's blood pressure
- hormone secretion. These hormones keep the wall of the uterus in 'pregnancy' state as the corpus luteum breaks down by the third month.

The placenta is lost as the *afterbirth* following birth of the fetus.

Wall of uterus: this is very muscular. At full term of the pregnancy a hormone (**oxytocin**) is secreted from the pituitary gland and makes this muscle contract in a series of waves to expel the fetus. The same hormone is used in the intravenous drip that induces birth when the pregnancy has gone on for too long.

Fetus: this develops from a **zygote**.

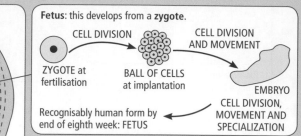

ZYGOTE at fertilisation — CELL DIVISION → BALL OF CELLS at implantation — CELL DIVISION AND MOVEMENT → EMBRYO

EMBRYO CELL DIVISION, MOVEMENT AND SPECIALIZATION → Recognisably human form by end of eighth week: FETUS

Amnion: the membrane that encloses the amniotic fluid. This is 'ruptured' just before birth.

Amniotic fluid: protects the fetus against
- mechanical shock
- drying out
- temperature fluctuations

Some of the fetal cells fall off into this fluid and can be collected by **amniocentesis**. The cells can be analysed to detect disease, genetic abnormalities and even the sex of the fetus.

UMBILICAL CORD: contains blood vessels which carry materials which will be/have been exchanged between mother and fetus. The cord connects the fetus to the placenta.

Mucus plug in cervix: protects fetus against possible infection. The plug is expelled just before birth.

Supplement   Supplement

Supplement

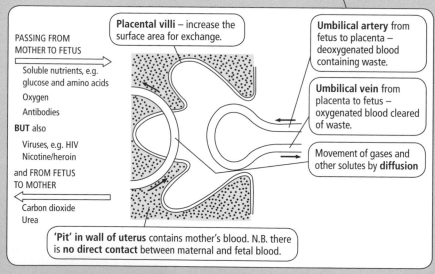

PASSING FROM MOTHER TO FETUS →

Soluble nutrients, e.g. glucose and amino acids

Oxygen

Antibodies

BUT also

Viruses, e.g. HIV
Nicotine/heroin

and FROM FETUS TO MOTHER ←

Carbon dioxide
Urea

Placental villi – increase the surface area for exchange.

Umbilical artery from fetus to placenta – deoxygenated blood containing waste.

Umbilical vein from placenta to fetus – oxygenated blood cleared of waste.

Movement of gases and other solutes by **diffusion**

'Pit' in wall of uterus contains mother's blood. N.B. there is **no direct contact** between maternal and fetal blood.

GROWTH AND DEVELOPMENT OF THE FETUS

MASS / g

Birth usually occurs at about this time

Fetus fully formed, even fingerprints!

Internal organs all present

Fetus now has a good chance of survival if born

AGE OF FETUS / weeks

- Complete period from fertilisation to birth = **gestation period**.
- Cell division converts single cell (zygote) to 30 million in a newborn baby.
- Most rapid growth from the twelfth week. As much as 1500 × gain in mass in 20 weeks.

1. The diagram below shows the reproductive system of a human male.

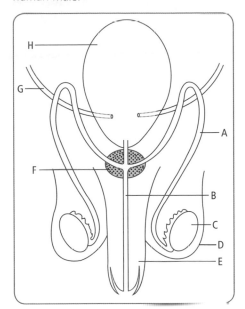

2. Arrange the following processes into the correct sequence necessary for the production of a human baby.
  a. ejaculation
  b. implantation
  c. birth
  d. fertilisation
  e. ovulation
  f. development   (6)

3. The diagram below shows part of the human female reproductive system.

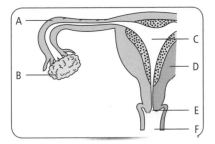

  a. Name the parts labelled **A**, **B**, **C**, **D** and **E**.   (5)
  b. Identify the part that produces the male gametes. (1)
  c. Identify the part that produces a liquid part of semen.   (1)
  d. Identify the part that produces testosterone.   (1)
  e. Identify the part that also carries urine.   (1)
  f. Identify the part that is cut during the procedure of vasectomy.   (1)

  a. Identify the parts labelled **A**, **B**, **C**, **D**, **E** and **F**.   (6)
  b. Make a simple copy of the diagram, and use an **X** to mark the place where fertilisation occurs and a **Y** to show where implantation occurs.   (2)
  c. Contraception prevents fertilisation occurring. Use the diagram to explain how **(i)** the diaphragm and **(ii)** the pill act as contraceptives.   (2)

4. During pregnancy, the developing fetus is attached to the mother's body through the placenta. The diagram shows the structure of this organ, and its attachment to the fetus.

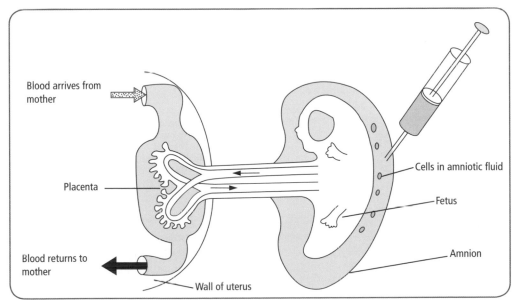

  a. i. Name **two** substances that pass from the fetus to the mother.   (2)
     ii. How is the structure of the placenta adapted to carry out this transfer?   (2)
    iii. Name **two** organs in the baby's body that take over the functions of the placenta once the baby is born. State the functions that these organs will perform.   (2)

b. A technique called amniocentesis can be used to check the genotype of the fetus. In this technique, a sample of amniotic fluid is removed using a long needle. The fluid contains cells shed from the body of the fetus. The cells can be stained, squashed and their chromosomes examined using a microscope.

   i. What is the function of amniotic fluid? (1)

   ii. Which condition would the baby have if the microscopist counted an extra chromosome 21? (1)

c. Scientists have been able to count the number of these abnormal chromosomes in mothers of different ages. The results are shown in the table.

	AGE OF MOTHER / YEARS					
	20–24	25–29	30–34	35–39	40–44	45–49
Frequency of abnormal fetuses per 1000 females	0.5	2.8	4.9	9.9	19.8	27.2

   i. Plot this information in the form of a bar graph. (3)

   ii. What is the percentage increase in abnormal fetuses between the age groups 30–34 and 45–49? Show your working. (2)

   iii. Women may be given hormone replacement therapy to delay the onset of menopause. Give **one** possible benefit and **one** possible disadvantage of this treatment. (2)

5. In women two hormones control ovulation (the release of eggs from the ovaries). The drawing shows a monitoring machine which women can use each morning to measure the amounts of the two hormones. A test stick is dipped in the woman's urine, then placed in a slot in the machine.

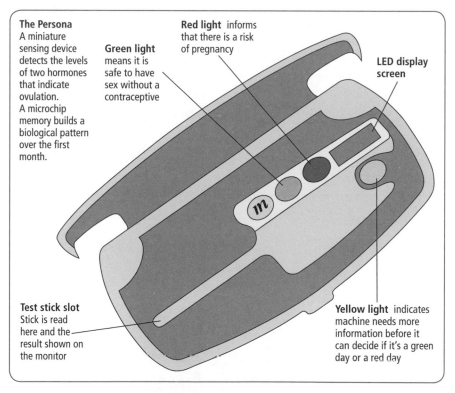

The Persona A miniature sensing device detects the levels of two hormones that indicate ovulation. A microchip memory builds a biological pattern over the first month.

Green light means it is safe to have sex without a contraceptive

Red light informs that there is a risk of pregnancy

LED display screen

Test stick slot Stick is read here and the result shown on the monitor

Yellow light indicates machine needs more information before it can decide if it's a green day or a red day

a. The machine monitors the levels of two hormones.

   i. What is a hormone? (1)

   ii. How are hormones transported around the body? (1)

b. A woman is unlikely to become pregnant if she has sex on the days when the machine shows a green light during the test. Use information from the drawing to suggest why. (1)

c. The two hormones detected by the machine are oestrogen and LH.
   Explain how oestrogen and LH are involved in the control of the menstrual cycle. (4)

d. Hormones can be used to control human fertility. Describe the benefits and problems that might arise from using hormones in this way. (4)

6. The drawing shows the human fetus at six different stages of development.

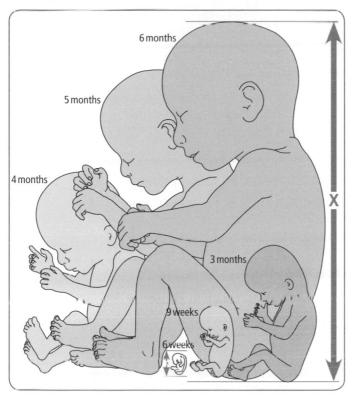

a. Apart from the increase in size, describe **one** change which occurs in the head and **one** change which occurs in the foot between the ages of **6 weeks** and **9 weeks**. **Use only features you can see in the drawing**. (2)

b. i. Measure the length **X** on the drawings of the fetus at
   A 6 weeks
   B 6 months (2)

   ii. How many times larger is the 6-month fetus than the 6-week fetus? (1)

REVISION SUMMARY: Match the term with the correct definition

	TERM		DESCRIPTION
A	Contraception	1	A developing mammal with recognizable internal and external organs
B	Embryo	2	Hormone that stimulates development of the corpus luteum
C	Fertilisation	3	Hormone promoting the development of a ripe ovum
D	Fetus	4	Hormone that keeps the lining of the uterus ready for implantation and pregnancy
E	FSH	5	Tube for the transfer of ova from ovary to the site of fertilisation
F	Gestation	6	The release of a female gamete
G	Implantation	7	The organ where materials are exchanged between maternal and fetal blood
H	LH	8	The fusion of male and female gametes to produce a zygote
I	Oestrogen	9	A developing human up to the eighth week of pregnancy
J	Ovary	10	Hormone that stimulates development of female secondary sexual characteristics
K	Oviduct	11	Hormone that stimulates production of sperm
L	Ovulation	12	Muscular chamber in which a young mammal develops
M	Placenta	13	Methods that prevent fertilisation and pregnancy
N	Progesterone	14	Structure linking a fetus to the placenta
O	Testes	15	Attachment of the zygote to the lining of the uterus
P	Testosterone	16	The birth canal
Q	Umbilical cord	17	The site for the production of sperm and male sex hormone
R	Uterus	18	The period between conception and birth
S	Vagina	19	Tube that delivers sperm to the penis
T	Vas deferens	20	Site of production of ova and female sex hormones

DNA, genes and chromosomes

Chromosome: a thread-like structure of DNA, carrying genetic information in the form of **genes**.

The nucleus of each human cell (mature red blood cells excepted) contains 46 chromosomes arranged as 23 **homologous pairs**. This is the diploid number for humans.

Because of a range of physical forces this ladder-like molecule 'twists' itself into a **double helix**.

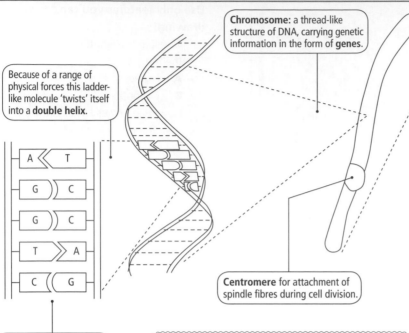

A	T
G	C
G	C
T	A
C	G

Centromere for attachment of spindle fibres during cell division.

Homologous pairs of chromosomes carry the same genes (e.g. the gene for eye colour) but may carry alternative forms, or **alleles** of the same gene (e.g. blue on one chromosome, brown on the other)

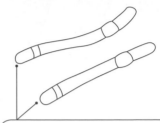

Gene: One gene is the length of DNA that codes for the production of one protein in the organism.

DNA is composed of chains of molecules called **nucleotide bases**. The sequence of bases on one chain codes for the sequence of amino acids in the protein. To protect the coded information from damage, the bases (represented here by the code letters A for Adenine, G for Guanine, C for Cytosine and T for Thymine) form up into double chains.

ALTERATIONS IN PHENOTYPE MAY RESULT FROM GENE MUTATIONS: gene mutations occur when some part of the base sequence in the DNA is altered. As a result a defective protein, or no protein at all, may be made.

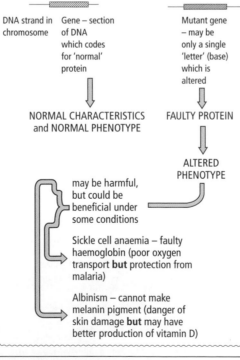

DNA strand in chromosome

Gene – section of DNA which codes for 'normal' protein

Mutant gene – may be only a single 'letter' (base) which is altered

NORMAL CHARACTERISTICS and NORMAL PHENOTYPE

FAULTY PROTEIN

ALTERED PHENOTYPE

may be harmful, but could be beneficial under some conditions

Sickle cell anaemia – faulty haemoglobin (poor oxygen transport **but** protection from malaria)

Albinism – cannot make melanin pigment (danger of skin damage **but** may have better production of vitamin D)

The complete set of alleles is the complete genetic make-up for the organism = **genotype**.

This genotype is largely responsible for all of the physical and other features of this organism (for example, hair and eye colour) = **phenotype**.

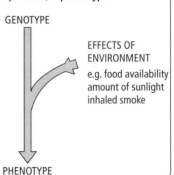

GENOTYPE

EFFECTS OF ENVIRONMENT e.g. food availability amount of sunlight inhaled smoke

PHENOTYPE

Protein synthesis
DNA/gene in nucleus
⟶ mRNA
• mRNA on ribosomes
• amino acids join to make protein.

RADIATION CAN INCREASE MUTATION RATES
Mutations **occur** spontaneously but there are factors that can increase the **rate** of mutation. Any factor that increases the rate of mutation is called a **mutagen**

→ some chemicals   e.g. agent orange (defoliant in Vietnam)

→ radiation   e.g. ultraviolet rays in sunlight, nuclear fall out

1. Chromosomes contain DNA. Each individual, with very few exceptions, has different DNA. The DNA can be broken up, and the pieces used to produce a 'genetic fingerprint'. The diagram shows the genetic fingerprints of blood found at a murder scene and from blood samples provided by five suspects.

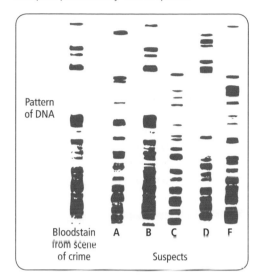

Fill in the missing words in the following sentences.

a. The chromosomes are found inside the ....
    of the cell. (1)
b. Genetic fingerprints must be made from the
    .... blood cells of a suspect. (1)
c   The murderer was most likely to be suspect
    .... (1)
d. Genetic fingerprints can only be identical if they
    come from .... (1)
e. Genetic fingerprinting can also be useful in
    conservation of endangered species. Scientists can
    examine the DNA fingerprint from a single hair
    of any animal. Suggest how this could be used to
    prevent dangerous inbreeding between rare animals
    kept in zoos. (2)

2. The diagram shows a section of a DNA molecule during replication.

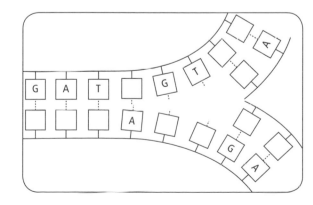

a. Copy the diagram, and complete it by writing
    in the correct letters in the empty boxes. (4)
b. Name the **four** molecules that these letters
    represent. (4)
c. DNA replication occurs during cell division. Why
    is replication important in the type of cell division
    called mitosis? (2)
d. What is the name of a section of DNA that codes
    for a particular protein? (1)
e. The table shows the codes for some of the amino
    acids found in human proteins.

ALANINE	GCC
CYSTEINE	ACA
GLYCINE	CCC
HISTIDINE	GTA
ISOLEUCINE	TAA
LYSINE	TTT
METHIONINE	TAC
PHENYLALANINE	AAA
PROLINE	CCG
TYROSINE	ATG
VALINE	CAT

The following is a sequence of bases in a section of DNA.

TAC – ATG – CCC – GCC – GTA

i.   Write out the sequence of amino acids coded
     by this DNA if it is read from left to right. (2)
ii. Sometimes a base letter is lost, perhaps as a
     result of exposure to radiation. Write out the
     new code sequence if the **ninth letter (a C) is
     lost** in this way. (2)
iii. Write out the sequence of amino acids if this
     altered DNA sequence is used. (2)
iv. What name is given to the changes in DNA
     that can affect the production of proteins? (1)
v.   Name **one** human disease that is caused by
     a faulty gene. (1)
vi. Explain why changes of this type can
     sometimes be an **advantage** to living
     organisms. (2)

Supplement

3. The diagram shows some of the steps in the manufacture of proteins.

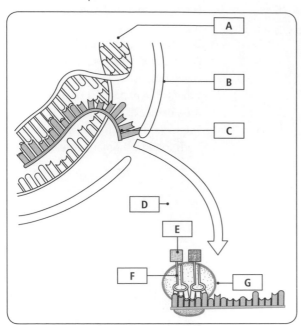

a. Match up the labels from this list with the letters **A–G** on the diagram. (7)
   TRANSFER RNA, CYTOPLASM, RIBOSOME, MESSENGER RNA, NUCLEAR MEMBRANE, DNA, AMINO ACID

b. Name a protein found in
   i. saliva
   ii. hair
   iii. red blood cells
   iv. stomach juices (4)

4. The sex of most animals is determined by their sex cells (gametes). It is these cells that combine to form a new organism. The diagram shows a male gamete from a human.

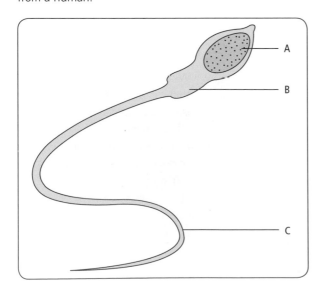

a. Name the type of cell division which produces gametes. (1)

b. State the major difference between a gamete and a normal cell. (1)

c. The part labelled **A** carries the instructions that will help to form the new organism. The instructions are carried as a chemical code. Name the chemical that carries these instructions. (1)

d. When a male and female gamete join together at fertilisation a zygote is formed. The zygote divides many times to become an embryo, and eventually a baby organism. During this division the 'information chemical' is copied accurately. The diagram shows a section of this chemical during this copying. Draw a completed version of the diagram by adding the correct code letters. (2)

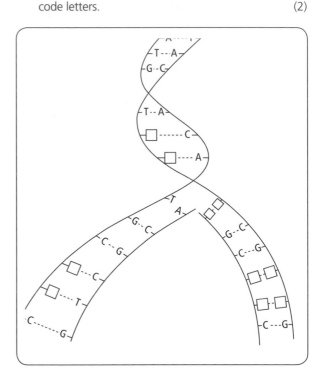

e. Sometimes the copying process goes wrong, and faulty information is made. What is the name given to the process causing 'faulty' copying? (1)

f. Name one environmental factor that might increase the risk of this faulty copying. (1)

g. Name one human disease caused by a fault of this type. (1)

5. a. The diagram shows the amount of DNA in a cell during cell division.

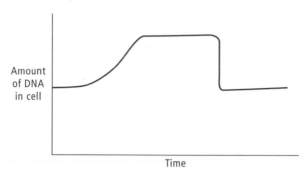

Amount of DNA in cell

Time

i. Name the type of cell division occurring in the cell. Explain how you arrived at your answer. (2)
ii. What is the function of DNA? (2)

b. The DNA content of a cell may be altered as a result of radiation. What name is given to this kind of change in a cell's DNA? (1)

c. The DNA content of a cell may also be changed deliberately, by genetic engineering. Use the following terms to explain this technique: BACTERIUM, CLONE, PLASMID, RESTRICTION ENZYME, LIGASE (5)

REVISION SUMMARY: Match the terms with the correct definition

A	Protein	1	DNA wound onto a protein scaffold
B	DNA	2	A single code letter in a DNA chain
C	Chromosome	3	Biological molecule responsible for a characteristic of an organism
D	Gene	4	Two chromosomes with the same genes in the same positions
E	Phenotype	5	This molecule is faulty in sufferers from sickle cell anaemia
F	Genotype	6	Ultraviolet, for example, can cause mutations
G	Nucleotide	7	The genetic material of all animals and plants
H	Homologous pair	8	An individual with a non-pigmented skin
I	Mutation	9	Alternative forms of the same gene
J	Haemoglobin	10	The total of the physical and other features of an organism
K	Albino	11	The arrangement of the DNA molecule
L	Radiation	12	A change in the type or amount of DNA
M	Alleles	13	A sequence of nucleotide bases coding for a single protein
N	Double helix	14	The complete set of alleles in a cell

DNA, GENES AND PROTEINS: Crossword

ACROSS:
2   Protein essential to the function of saliva
4   Type of molecule responsible for a characteristic in a cell or organism
5   The total of all the genetic material in a nucleus
7   Some factor which can change DNA
10 A long strand of genes
13 Type of RNA which carries genetic information from the nucleus to the cytoplasm
14 The 'letters' that make up the code in DNA
15 Sub-unit of a protein
16 Technical name for the copying of DNA
18 The total of all the characteristics of an organism
19 Important protein for oxygen transport
20 A section of DNA that codes for a single protein

DOWN:
1   This type of cell has DNA but not inside a nucleus
3   A change to the type or amount of DNA
6   Where to look for DNA in a plant, animal or fungal cell
8   Process that 'reads' the message from the nucleus and produces a protein
9   The organelles in the cytoplasm where proteins are produced
11 Ultraviolet, for example – an example of 7 across
12 The arrangement of the DNA chains
17 An alternative version of 20 across

HUMAN LIFE CYCLE

Male parent (2n = 46) — MEIOSIS — sperm (n = 23), sperm (n = 23)

Female parent (2n = 46) — MEIOSIS — egg (n = 23), egg (n = 23)

Fertilisation – return to normal (diploid) number of chromosomes by the fusion of the two gametes.

zygote (2n = 46)

Mitosis – provides more cells as 'building blocks' of new organism

New adult (2n = 46)

Meiosis is necessary to halve the chromosome number from diploid (2n) to haploid (n).

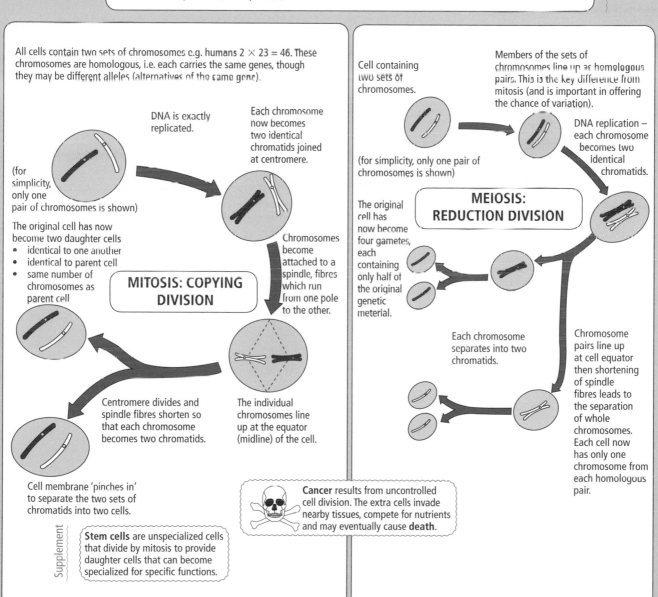

All cells contain two sets of chromosomes e.g. humans 2 × 23 = 46. These chromosomes are homologous, i.e. each carries the same genes, though they may be different alleles (alternatives of the same gene).

DNA is exactly replicated.

Each chromosome now becomes two identical chromatids joined at centromere.

(for simplicity, only one pair of chromosomes is shown)

The original cell has now become two daughter cells
• identical to one another
• identical to parent cell
• same number of chromosomes as parent cell

Chromosomes become attached to a spindle, fibres which run from one pole to the other.

MITOSIS: COPYING DIVISION

Centromere divides and spindle fibres shorten so that each chromosome becomes two chromatids.

The individual chromosomes line up at the equator (midline) of the cell.

Cell membrane 'pinches in' to separate the two sets of chromatids into two cells.

Cell containing two sets of chromosomes.

Members of the sets of chromosomes line up as homologous pairs. This is the key difference from mitosis (and is important in offering the chance of variation).

DNA replication – each chromosome becomes two identical chromatids.

(for simplicity, only one pair of chromosomes is shown)

The original cell has now become four gametes, each containing only half of the original genetic material.

MEIOSIS: REDUCTION DIVISION

Each chromosome separates into two chromatids.

Chromosome pairs line up at cell equator then shortening of spindle fibres leads to the separation of whole chromosomes. Each cell now has only one chromosome from each homologous pair.

Cancer results from uncontrolled cell division. The extra cells invade nearby tissues, compete for nutrients and may eventually cause **death**.

Stem cells are unspecialized cells that divide by mitosis to provide daughter cells that can become specialized for specific functions.

1. a. What is a chromosome?
   b. How many chromosomes are there in
       i.   a skin cell from a human male?
       ii. a human egg cell?
       iii. a red blood cell from a human female?
       iv. a red blood cell from a human male?   (4)
   c. Name the process in which haploid cells are
       formed from diploid cells.   (1)
   d. Name the process which provides new cells for
       the growth of a young mammal.   (1)

2. This diagram represents the life cycle of a mammal.

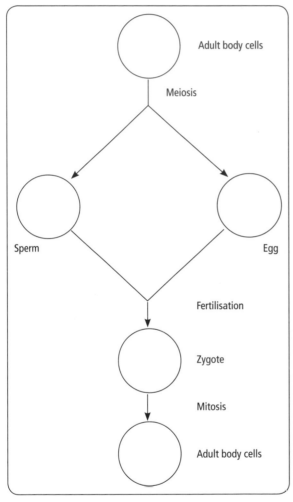

   a. Copy and complete the diagram by writing in the
       circles the numbers of chromosomes found in
       the nuclei of these cells if the mammal concerned
       was a human.   (5)
   b. In what way is this diagram misleading?   (1)
   c. Where, exactly, does meiosis occur in a mammal? (1)
   d. Why is it necessary for gametes to be formed by
       meiosis?   (2)

Supplement

3. It is possible to observe the process of mitosis in cells
    that are actively dividing. The tips of roots and shoots
    have many dividing cells when a plant is increasing in
    length.

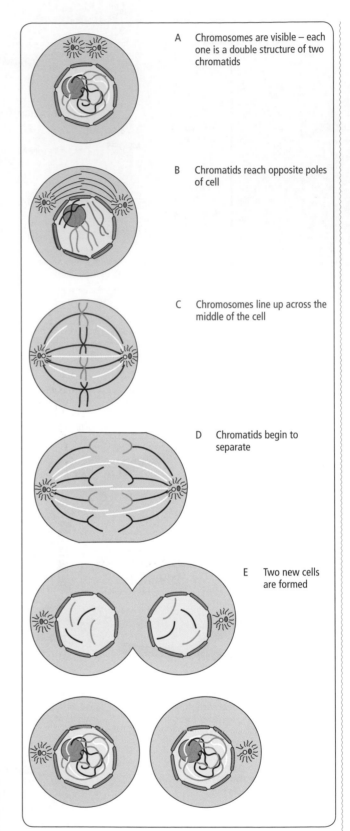

A Chromosomes are visible – each
   one is a double structure of two
   chromatids

B Chromatids reach opposite poles
   of cell

C Chromosomes line up across the
   middle of the cell

D Chromatids begin to
   separate

E Two new cells
   are formed

A student was able to cut off the top 5 mm of a plant
shoot. She then softened it by warming it in a dilute
solution of hydrochloric acid, and squashed the tip onto a
microscope slide. The squashed cells were treated with a
red dye, and then observed under a light microscope.
a. Why did she soften the tip?   (1)
b. Why did she squash the tip?   (1)

c. Why is red dye added to the preparation? (1)
d. The student took some photographs of the dividing cells. These are shown in the figure on the previous page. What would be the sequence of these cells as they divided? (4)

e. Make a labelled drawing of one chromosome from figure **A**. (2)
f. How many chromatids are present in figure **B**? (1)
g. How many chromosomes would be present in a body cell from this organism? (1)

4. Read the following article, and then answer the questions that follow it.

WHAT CAUSES CANCERS?

The rate of cell division by mitosis is normally very strictly controlled. The control normally involves cells touching one another as they fill up the available space – once the space is properly filled the cells 'switch off' their process of division. This is called **contact inhibition** and involves special sensitive proteins on the cell surface membrane. Cell division may get out of control for a number of reasons:

- The cell may not have the gene to produce the correct cell surface protein – this is a genetic tendency, and explains why some types of cancer (breast cancer is an example) tend to run in families.
- The protein may be inactivated in some way. This usually occurs due to an **environmental** factor, and explains why certain lifestyles or occupations carry a higher risk of cancer. Smoking of tobacco, for example, increases the risk of cancer of the lung and heavy consumption of alcohol increases the risk of cancer of the oesophagus.

Scientists believe that most types of cancer only develop when several factors are present. For example, a person might be carrying a gene that makes their lung cells more likely to divide, but the presence of chemicals in tobacco smoke is necessary to make the gene run out of control. Any factor in the environment that causes cancer to develop is called a **carcinogen** (literally a 'cancer maker') – the involvement of external factors in the development of cancer of the lung, for example, has been very widely studied.

THE TREATMENT OF CANCER

If a cancer is detected early enough it can often be treated. Hospitals run **screening programmes** to try to detect some cancers, such as breast cancer, cervical cancer and lung cancer, while they are still treatable. Doctors urge their patients to visit the surgery if they have any concern about their health, especially sudden weight loss or unexplained bleeding. Treatment always involves exerting some control of the cancer cells again, either by surgery, chemotherapy or radiotherapy.

'Cancer' is actually a very general term. There are many forms of cancer, and many tissues may develop tumours. Since cancer results from cell division that has gone out of control, the disease is most likely to occur in tissues that are normally dividing quite frequently. For this reason cancer is particularly common in the sites in the body where mitosis is occurring.

---

a. What are the **two** general causes of cancer? (2)
b. How do dividing cells 'decide' when to stop their division? (1)
c. Which type of cell division, if uncontrolled, can lead to cancer? (1)
d. Suggest **one** site in the body where this type of cell division is occurring. (1)
e. What name is given to a factor that causes cells to divide in an uncontrolled way? (1)
f. What are the **three** general methods of treating cancer? (3)

REVISION SUMMARY: Match the terms with the correct definition

	TERM			DESCRIPTION
A	Mitosis		1	The number of pairs of chromosomes in a normal body cell
B	Meiosis		2	A copied chromosome
C	Fertilisation		3	Structure that holds the chromosome to the spindle as cell division proceeds
D	Zygote		4	A sex cell – the product of meiosis
E	Haploid		5	Reduction division, which produces haploid daughter cells from a diploid parent cell
F	Diploid		6	The diploid product of fertilisation
G	Chromatid		7	Copying division, essential for growth, repair and asexual reproduction
H	Spindle		8	The fusion of gametes as part of sexual reproduction
I	Centromere		9	The total number of chromosomes in a normal body cell
J	Gamete		10	Framework for moving chromosomes or chromatids to the poles during cell division

Chapter 30:
Patterns of inheritance

Cystic fibrosis in humans is an example of monohybrid inheritance

SOME IMPORTANT DEFINITIONS

INHERITANCE – transmission of genetic information from generation to generation

HOMOZYGOUS – having two identical alleles for a particular gene e.g. NN or nn

HETEROZYGOUS – having two different alleles for a particular gene e.g. Nn

DOMINANT – an allele which is expressed if it is present. Usually given a capital letter e.g. N

RECESSIVE – an allele which is expressed only when there is no dominant allele of the gene present. Usually given a small letter e.g. n

GENOTYPE – the genetic makeup of an organism in terms of the alleles present

PHENOTYPE – the features (characteristics) of an organism

Cystic fibrosis (CF) results from an imbalance of chloride ions across the membranes of cells lining some of the main passageways of the body, causing dangerous buildup of mucus.

It is the most common inherited fatal disease – about 1 in 20 white Europeans is a carrier of the mutant allele.

The condition is caused by a recessive allele – the 'non-disease' state is dominant.

So normal = N, with condition = n

If both parents are carriers (i.e. heterozygous)

PARENTAL GENERATION   Nn × Nn

GAMETES (N) (n)   (N) (n)

A Punnett square can be used to predict the possible combinations of alleles in the zygote

Gametes from father ♂	(N)	(n)
♀ (N)	NN	Nn
(n)	Nn	nn

Gametes from mother

Phenotype: no cystic fibrosis
Genotype: normal

Phenotype: no cystic fibrosis
Genotype: carrier of mutant allele

Phenotype: has cystic fibrosis
Genotype: homozygous for mutant allele

3 no cystic fibrosis: 1 with cystic fibrosis

N.B. The 3:1 ratio is only approximate unless the number of offspring is very large (unlikely in humans), because:

1. alleles may not be distributed between viable gametes in equal numbers
2. fusion of gametes is completely random. It is a matter of chance whether one male gamete fuses with a particular female gamete.

If one parent's genotype is homozygous normal, the other parent's is homozygous mutant, then theoretically:

The homozygous normal parent is represented as NN, the parent with homozygous mutant genotype as nn, since in this case normal is dominant to mutant.

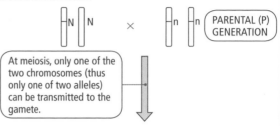

N N   ×   n n   PARENTAL (P) GENERATION

At meiosis, only one of the two chromosomes (thus only one of two alleles) can be transmitted to the gamete.

(N)   (n) GAMETES

At fertilisation, fusion of gametes to form a zygote restores the diploid number.

F₁ GENERATION

Nn

The allele that 'shows up' in this heterozygote is DOMINANT, the other allele (which remains 'hidden') is RECESSIVE

This individual is **genotypically** heterozygous, but **phenotypically** normal i.e. does not have the disease but is a **carrier** of the mutant allele for cystic fibrosis.

SOLVING INHERITANCE PROBLEMS

For example, two parents **without** CF have a child **with** CF. Work out the genotypes of both parents and their child.

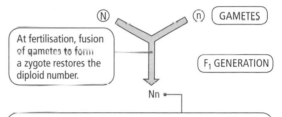

🔑 THE KEY! The only genotype you can be sure of is the homozygous recessive, i.e. CF sufferer **must** be nn…

…so non-CF parents **must** be Nn (since each must hand on one n allele to the offspring).

Nn   Nn

CF
nn

TEST CROSS can show if an individual without the condition is homozygous or heterozygous

Homozygous normal × homozygous recessive

NN × nn → all 'normal' offspring

Heterozygous normal × homozygous recessive

Nn × nn → 1:1 normal : recessive

N.B. You can't do test crosses on humans!

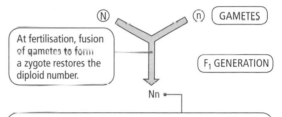 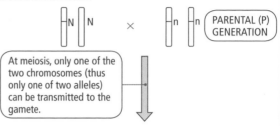

Supplement

Co-dominance sex linkage and the inheritance of sex

Supplement

CODOMINANCE

Some genes have more than two alleles. For example, the gene controlling the human ABO blood groups has three alleles, given the symbols I^A, I^B and I^O. Neither of the I^A and I^B alleles is dominant to the other, although they are both dominant to I^O. This is called **codominance**. It results in an extra phenotype when both alleles are present together. The genotypes and phenotypes are shown in the table.

Genotype	Phenotype
$I^A I^A$ or $I^A I^O$	Blood group A
$I^B I^B$ or $I^B I^O$	Blood group B
$I^A I^B$	Blood group AB
$I^O I^O$	Blood group O

The human blood groups are easily detected by a simple test on a blood sample.

This can be interesting when a woman can't be sure of the father of a child. For example, could a type O man be the father of a type AB child?

♂ gametes	I^O	I^O
♀ gametes: no ♀ could provide both I^A and I^B	Child $I^A I^B$	

NO! This man could NOT be the father!

A KARYOTYPE is obtained by rearranging photographs of stained chromosomes observed during mitosis. Such a karyotype indicates that

① the chromosomes are arranged in **homologous pairs.** In humans there are 23 pairs and we say that the **diploid number** is 46 ($2n = 46 = 2 \times 23$).

② whereas females have 22 pairs + XX in the karyotype, males have 22 pairs + XY, i.e. the 23rd 'pair' would **not** be two copies of the X chromosome.

The Y chromosome is so small that there is little room for any genes other than those responsible for 'maleness', but the X chromosome can carry some genes as well as those for 'femaleness' – these additional genes are **X-linked** (usually described as **sex-linked**).

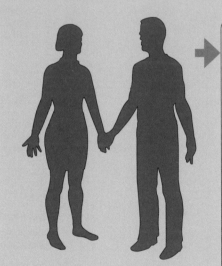

INHERITANCE OF SEX is a special form of monohybrid inheritance

♂ (XY)   ♀ (XX)

GAMETES Ⓧ Ⓨ   Ⓧ Ⓧ

F_1 generation: sex of offspring can be determined from a Punnett square

♂ GAMETES / ♀ GAMETES	X	Y
X	XX (female)	XY (male)
X	XX (female)	XY (male)

- theoretically there should be 1:1 ratio of male to female
- the male gamete determines the sex of the offspring

SEX-LINKED INHERITANCE

If the X chromosome from the mother carries a gene for an inherited disease like colour blindness (X^n) then her male children may be colour blind. A female child would be unlikely to be colour blind because a normal vision gene (X^N) on her inherited X chromosome would be dominant. However, she could pass the gene on to her sons – she is a **carrier**.

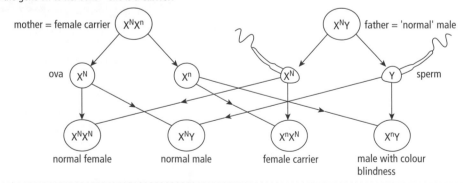

Supplement

1. Choose words from the following list to match the definitions below.

   GENOTYPE, ALLELE, MENDEL, DARWIN, GENE, PHENOTYPE, HETEROZYGOTE, DOMINANT, HOMOZYGOTE

   a. The external appearance of an organism.
   b. Alternative form of a gene.
   c. Studied genetics in garden peas.
   d. Section of DNA coding for a single characteristic.
   e. The set of genes in the nucleus of any individual.
   f. A nucleus carrying both alternative alleles of a gene.
   g. An allele that determines the appearance of a heterozygote. (7)

2. Cystic fibrosis is one of the most common inherited diseases in humans. The disease is caused by the inheritance of a recessive allele – about 1 in 20 white-skinned individuals are heterozygous for this allele. Heterozygotes do not show symptoms of the disease, but are carriers of the allele.

   a. Copy and complete the following genetic diagram to show the possible inheritance of cystic fibrosis by children from two carrier parents. (Let R = normal allele; let r = allele for cystic fibrosis) (7)

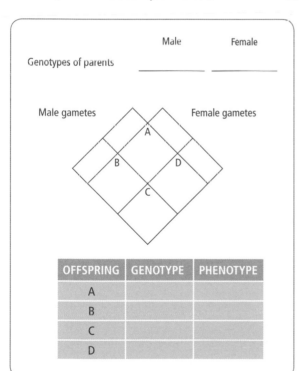

OFFSPRING	GENOTYPE	PHENOTYPE
A		
B		
C		
D		

   b. Two carrier parents have two children, neither of whom show any symptoms of cystic fibrosis. What is the chance that a third child will have cystic fibrosis? (2)

   c. It is possible to use a gene probe to detect the cystic fibrosis allele – the probe is able to bind onto this allele in a sample of a person's DNA. The DNA is collected from blood. Which type of blood cell would supply the DNA for the test? Explain your answer. (2)

3. Haemophilia is a sex-linked inherited characteristic. The allele (n) for haemophilia is recessive to the allele (N) for normal blood clotting. The gene responsible for blood clotting is carried only on the X chromosome. The phenotype of an organism is the appearance or characteristic that results from the inheritance of a particular combination of alleles.

   a. Describe the phenotypes of individuals with the following alleles. (4)

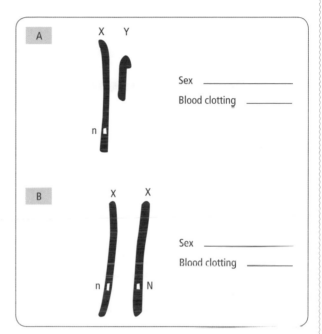

   b. Use genetic symbols to explain how two non-haemophiliac parents could have a haemophiliac son. (4)

   c. Explain
      i. why haemophiliac females are less common than haemophiliac males
      ii. why haemophilia is especially dangerous for girls. (2)

4. Sickle cell anaemia is a disease in which people produce abnormal haemoglobin in their red blood cells. This abnormal haemoglobin makes the cells take on a sickle shape, especially if the person exercises and produces lactic acid. The sickle cells do not carry oxygen as well as normal cells do.
   Let N = the allele for 'normal' haemoglobin; let S = the allele for 'sickle' haemoglobin.

   a. Copy and complete the genetic diagram to show the results of a cross between two carrier parents. (4)

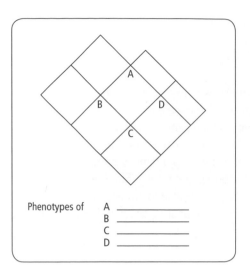

Phenotypes of A _____
                B _____
                C _____
                D _____

b. Explain why a person with sickle cell anaemia would often feel tired. (1)

c. People who are carriers of this condition are resistant to a blood-borne disease transmitted by mosquitoes. What is the name of this disease? (1)

d. Name the process in which a 'normal' allele might be converted to a 'sickle' allele. (1)

e. Name **one** environmental factor that might cause a change of this type. (1)

5. Cystic fibrosis is a serious condition in humans. Many of the body tubes become blocked with a very sticky mucus. Bacteria invade the mucus and cause serious infections, and tubes and passageways blocked by the mucus do not allow the passage of important juices. Organs most affected by the condition are the lungs, pancreas and reproductive organs. The disease is inherited, and involves a single gene with two alleles.

a. Two parents who showed no symptoms of the condition had two children. The second child had cystic fibrosis although the other child was healthy.
   i. Is the allele that causes cystic fibrosis dominant or recessive? Explain your answer. (2)
   ii. The parents decide to have another child. What is the chance that the third child will have cystic fibrosis? Use a genetic diagram to explain your answer. (4)

b. The condition is treated by giving patients antibiotics and capsules of enzymes. Explain why this treatment is effective. (2)

c. Read this extract from a scientific journal.

> Medical researchers have developed a method for correcting cystic fibrosis. The treatment involves introducing a normal allele of the gene for a salt-transporting protein into the lungs of patients suffering from this condition. The normal gene is inserted into the coat of a virus similar to the one that causes the common cold. The virus's own genetic material is removed. The patient inhales the genetically engineered virus, the lung cells are invaded by the virus and stop producing the sticky mucus.

   i. Draw a series of diagrams to explain how the virus could be genetically engineered to contain the 'normal' gene. (4)
   ii. Why are some people very much against the use of genetically engineered organisms? (3)

REVISION SUMMARY: Fill in the gaps

Use terms from the following list to complete the paragraphs below. You may use each term once, more than once or not at all.

GENES, NUCLEUS, DARWIN, MENDEL, TALL, DWARF, 3:1, DNA, 1:1, CHROMOSOMES, STAMENS, INSECTS, GARDEN PEA, STIGMA, HETEROZYGOUS, RECESSIVE, DOMINANT, HOMOZYGOUS, HEIGHT

An Austrian monk called .... carried out a series of important genetic investigations on plants of the .... To control the inheritance of characteristics he removed the .... from some plants and dusted their pollen onto the .... of other plants of the same species. He then covered these flowers in muslin bags so that they could not be pollinated by ....

In one experiment he crossed .... plants with dwarf plants to investigate the inheritance of .... in this species. When he used pure-breeding (........................) parent plants, all of the offspring were .... − this is the .... characteristic. When these offspring were self-fertilised they produced tall and dwarf plants in a ratio of ....

Modern work in biochemistry was unknown to ...., and he knew nothing of the genetics familiar to us. For example, he was unaware that characteristics are controlled by short sections of .... carried as .... on .... He did, however, have the use of a microscope and did appreciate that the 'factors of inheritance' were carried in the .... of the cell. (16)

Variation and natural selection may lead to evolution of species

Environmental resistance describes the factors, e.g. food availability, that limit the growth of populations. Animals and plants that are best **adapted** to their environment suffer from less environmental resistance. **Adaptations** are often structural, but they can also be biochemical or behavioural.

These adaptations arise because there is **variation** between the different members of a population. Charles Darwin studied many examples of these adaptations, and published his conclusions in *The Origin of Species by Means of Natural Selection*. His observations can be summarized under a number of headings.

Over-production: All organisms produce more offspring than can possibly survive, but populations remain relatively stable.

e.g. a female peppered moth may lay 500 eggs, but the moth population does not increase by 25 000 %!

Struggle for existence: Organisms experience environmental resistance, i.e. they compete for the limited resources within the environment.

e.g. several moths may try to feed on the same nectar-producing flower.

Variation: Within the population there may be some characteristics that make the organisms that possess them more suited for this severe competition.

e.g. moths might be stronger fliers, have better feeding mouthparts, be better camouflaged while resting or be less affected by rain.

Survival of the fittest: Individuals that are most successful in the struggle for existence (i.e. are the best suited / adapted to their environment) will survive more easily than those without these advantages.

e.g. peppered moths: dark coloured moths resting on soot-covered tree trunks will be less likely to be captured by predators.

Advantageous characteristics are passed on to offspring: The well adapted individuals breed more successfully than those that are less well adapted – they pass on their genes to the next generation. This process is called **natural selection**.

e.g. dark coloured moth parents will produce dark coloured moth offspring.

HOW GENETIC VARIATION MAY ARISE

- both chromosomal and gene **mutation** provide **raw material for variation**
- meiosis and sexual reproduction then **re-arrange this raw material** to provide **many new genotypes**, by (a) crossing over, (b) independent assortment and (c) fertilisation

TYPES OF VARIATION

1 DISCONTINUOUS:
- caused by GENES only
- limited number of phenotypes with no intermediates

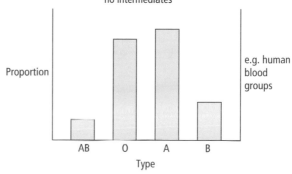

e.g. human blood groups

2 CONTINUOUS:
- caused by GENES and EFFECTS OF ENVIRONMENT
- range of phenotypes between the two extremes

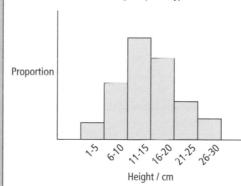

e.g. height of wheat plants – affected by soil nutrients, light intensity and water availability

---

SUMMARY:

VARIATION $\xrightarrow[\text{SELECTION}]{\text{NATURAL}}$ ADAPTATION $\xrightarrow[\text{SELECTION}]{\text{MORE NATURAL}}$ NEW SPECIES

GENES AND NATURAL SELECTION

Individuals' heterozygous for sickle cell allele $Hb^S Hb^A$ are resistant to malaria, so survive and reproduce in regions where malaria is common.

ANTIBIOTIC RESISTANT BACTERIA can result from natural selection (the presence of the antibiotic) acting on a population of bacteria. MRSA is an important example.

Supplement

Supplement

Selective breeding and cloning reduce variation

Selective breeding and cloning are methods we can use to produce plants and animals better suited to our purposes.

SELECTIVE BREEDING

For hundreds of years farmers have been breeding animals and plants to give greater productivity. By selecting parents with desired characteristics and mating them, there is a better chance of the offspring having those characteristics. For example, if a champion stallion (male horse) is mated with a champion mare (female horse) the foals (young horses) are likely to be fast runners.

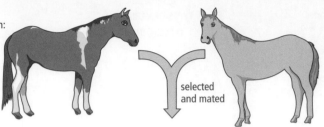

champion stallion:
fast
large muscles
healthy
good stamina

selected
and mated

champion mare:
fast
large muscles
healthy
good stamina

offspring likely, but not certain, to inherit parents' characteristics

The basic plan
Humans select individuals with desirable characteristics

These individuals are crossed to produce the next generation

Offspring with desirable characteristics are selected, then used for further breeding

Selective breeding has been very successful but it is slow and unpredictable. Genetic engineering (see p. 140) is quicker and more reliable.

CLONING

Cloning is the artificial production of many identical plants or animals. One method of cloning animals fertilises an egg outside of the female (*in vitro fertilisation*).

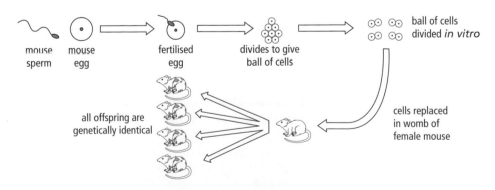

mouse sperm

mouse egg

fertilised egg

divides to give ball of cells

ball of cells divided *in vitro*

cells replaced in womb of female mouse

all offspring are genetically identical

Uses of cloning:
1 Making identical plants and animals for use in controlled scientific research.
2 Making identical copies of animals and plants which are good for agriculture.
3 Making large numbers of rare and expensive plants, e.g. propagating orchids.
4 Increasing the chances of having a baby for couples who have difficulty conceiving.
5 Screening for inherited disease.

1. The table shows some examples of variation In living organisms. For each of the examples, indicate whether the variation results from **artificial** (A) or from **natural** (N) selection.

NUMBER	VARIATION OBSERVED	ARTIFICIAL OR NATURAL
1	Cattle bred for high milk yield	
2	Development of wheat strains resistant to infection by fungi	
3	Sheep bred for thick wool	
4	The development of thick fur by polar bears	
5	Bacteria developing antibiotic resistance	
6	Banding on snail shells offers camouflage from song thrushes	
7	Birch trees growing on mine spoil heaps with a high lead content	
8	Members of the *Brassica* (cabbage) group chosen as sprouts because of many small side buds	

2. A group of students carried out an investigation into the diameter of the bases of limpet shells from the same rocky shore. They presented their results in the two histograms shown below:

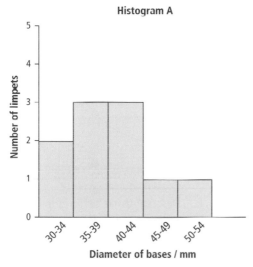

Histogram A

Number of limpets (y-axis) vs *Diameter of bases / mm* (x-axis: 30-34, 35-39, 40-44, 45-49, 50-54)

Histogram B

Number of limpets (y-axis) vs *Diameter of bases / mm* (x-axis: 30-34, 35-39, 40-44, 45-49, 50-54)

a. How many limpet shells were measured to produce the data shown in the two histograms? (2)

b. Which of the two histograms, A or B, would give the more reliable mean diameter of the base of the limpet shells? Explain your answer. (2)

The limpet uses its muscular foot to hold onto rocks. The diameter of the base of the shell is a very good measure of the size of the foot. The students carried out a further investigation into base diameter of limpets from two different parts of the shore.

	AREA 1	AREA 2
Mean diameter of shell bases / mm	39	47

c. Which of the two areas, 1 or 2, had the biggest waves? Explain your answer. (2)

d. What type of variation is shown by the diameter of the bases of the limpet shells? (1)

e. Give an example of the same type of variation in humans. (1)

3. Snails of one species vary as shown.

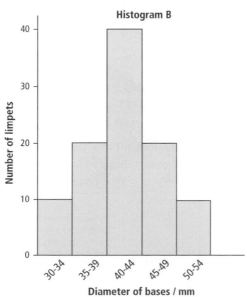

A scientist made the following observations on a group of snails.

1. A large number of snails are born.
2. Some snails have no bands on their shells. Others have bands on their shells.

3. Snails without bands on their shells tend to produce young without bands on their shells.
4. In hot conditions, snails with bands on their shells are more likely to die of heat shock.

The table gives **three** statements about the theory of evolution by natural selection.

a. Write the correct numbers in the table so that the observations made by the scientist match the following statements about evolution. (3)

STATEMENT	NUMBER OF MATCHING OBSERVATIONS
Some variations are inherited	
All populations show variation	
Natural populations over-produce	

b. A scientist did an investigation in which he used a special paint to mark snail shells. The paint was sensitive to daylight. He put a small spot of this paint on the top of the snails' shell. He used 200 snails with banded shells, and 200 snails with unbanded shells. He put the snails into ten cages in areas planted with grass and nettles. Sixty days later he looked at how much the paint had faded.

The table shows his results. The higher the figure, the more faded the paint is.

	BANDED SNAIL SHELLS	UNBANDED SNAIL SHELLS
Average fading of paint spot on shell	4.55	3.85

   i. Why did he not have to worry about the paint spots making the snails easy to see by predators? (1)
   ii. How was he able to work out which type of snail stayed longer in the sunlight. (1)
   iii. Why would putting a spot of paint in the same place on each snail's shell help make it a fair test? (1)
   iv. Suggest why dark banded snails have an advantage over unbanded snails in cold and shady places. (2)

4. The diagram shows a plant that is well adapted to life in a particular environment.

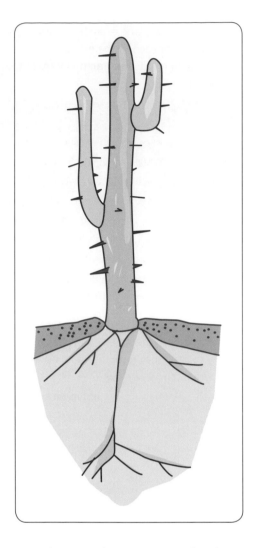

a. What type of environment is this plant suited to? Explain your answer. (2)
b. Describe **three** features that suit the plant to its environment. (3)
c. Match the letters and the numbers to explain how the mammal shown below is well adapted to its environment. (6)

A	Stored fat can be respired to release water
B	Enclosed in membrane to limit evaporation
C	Large area makes movement more efficient
D	Can be closed to prevent entry of sand
E	Reduced fur thickness aids heat conduction
F	Increased fur thickness limits heat loss

REVISION SUMMARY: Fill in the gaps

Complete the following paragraphs. Use words and phrases from the following list – you may use each word once, more than once or not at all.

GENES, FERTILISATION, EFFECTS OF ENVIRONMENT, PHENOTYPE, MUTATION, DISCONTINUOUS, CONTINUOUS, NATURAL SELECTION, EVOLUTION, CROSSING OVER, BLOOD GROUPING, ENVIRONMENTAL, ATMOSPHERIC, INDEPENDENT ASSORTMENT, GENOTYPE, BODY MASS, NUTRIENTS

There are two kinds of variation – the first is ...., which shows clear cut separation between groups showing this variation (........................, for example). The second is ...., in which there are many intermediate forms between the extremes of the characteristic. A clear example of this second type is ....

.... is the result of .... alone, while .... is also affected by .... factors. The sum of the genes that an organism contains is called its .... and the total of all its observable and measurable characteristics is called the .... The two are related in a simple equation .... equals .... plus ....

Variation that can be inherited results from ...., which provides the raw material that can be rearranged by .... and .... during meiosis, and the process of .... when male and female gametes fuse.

Variation provides the raw material for .... Organisms may gain an advantage in the struggle for existence, and .... can 'pick out' the organisms with such an advantage. These organisms may then reproduce and pass on their .... to their offspring.   (20)

Ecology is the study of living organisms in relation to their environment.

THIS MAY INVOLVE

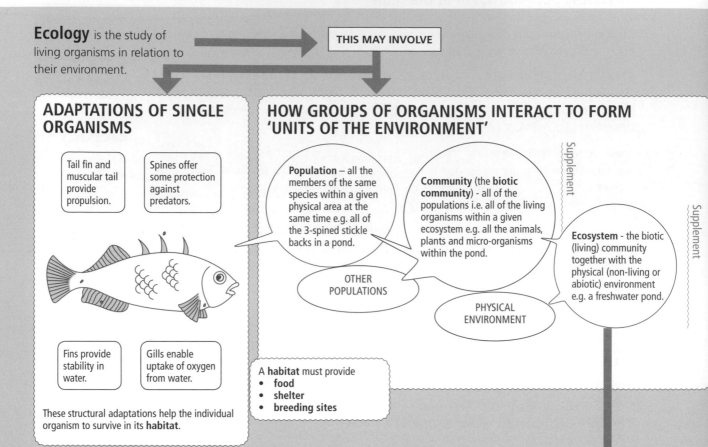

ADAPTATIONS OF SINGLE ORGANISMS

Tail fin and muscular tail provide propulsion.

Spines offer some protection against predators.

Fins provide stability in water.

Gills enable uptake of oxygen from water.

These structural adaptations help the individual organism to survive in its **habitat**.

A **habitat** must provide
- **food**
- **shelter**
- **breeding sites**

HOW GROUPS OF ORGANISMS INTERACT TO FORM 'UNITS OF THE ENVIRONMENT'

Population – all the members of the same species within a given physical area at the same time e.g. all of the 3-spined stickle backs in a pond.

OTHER POPULATIONS

Community (the **biotic community**) - all of the populations i.e. all of the living organisms within a given ecosystem e.g. all the animals, plants and micro-organisms within the pond.

PHYSICAL ENVIRONMENT

Ecosystem - the biotic (living) community together with the physical (non-living or abiotic) environment e.g. a freshwater pond.

Supplement

Supplement

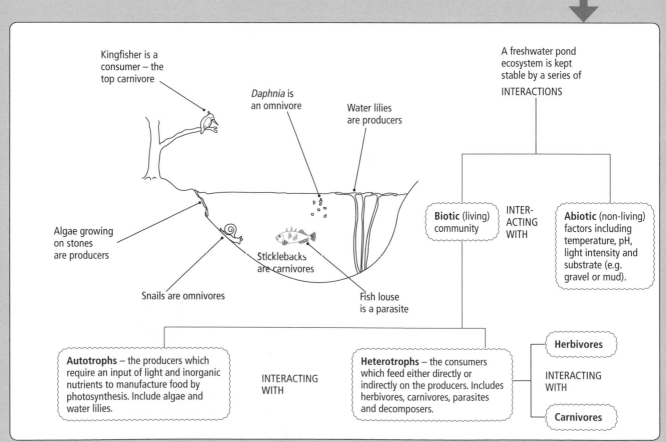

Kingfisher is a consumer – the top carnivore

Daphnia is an omnivore

Water lilies are producers

Algae growing on stones are producers

Snails are omnivores

Sticklebacks are carnivores

Fish louse is a parasite

A freshwater pond ecosystem is kept stable by a series of INTERACTIONS

Biotic (living) community

INTER-ACTING WITH

Abiotic (non-living) factors including temperature, pH, light intensity and substrate (e.g. gravel or mud).

Autotrophs – the producers which require an input of light and inorganic nutrients to manufacture food by photosynthesis. Include algae and water lilies.

INTERACTING WITH

Heterotrophs – the consumers which feed either directly or indirectly on the producers. Includes herbivores, carnivores, parasites and decomposers.

INTERACTING WITH

Herbivores

Carnivores

Food chains and food webs show energy flow through an ecosystem.

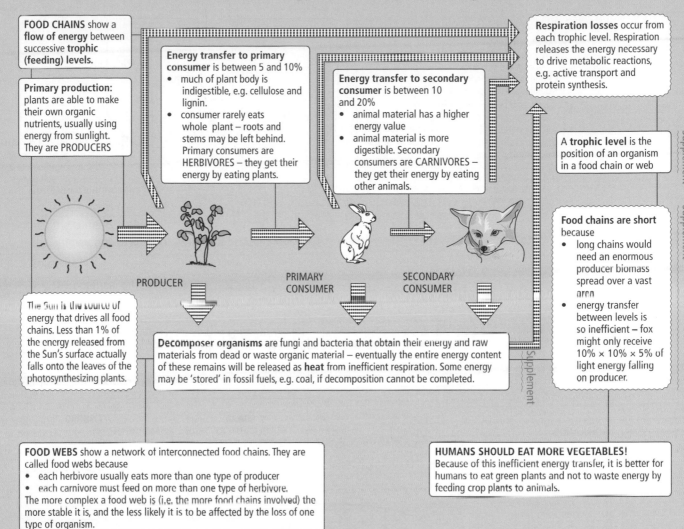

FOOD CHAINS show a flow of energy between successive **trophic (feeding) levels.**

Primary production: plants are able to make their own organic nutrients, usually using energy from sunlight. They are PRODUCERS

Energy transfer to primary consumer is between 5 and 10%
- much of plant body is indigestible, e.g. cellulose and lignin.
- consumer rarely eats whole plant – roots and stems may be left behind. Primary consumers are HERBIVORES – they get their energy by eating plants.

Energy transfer to secondary consumer is between 10 and 20%
- animal material has a higher energy value
- animal material is more digestible. Secondary consumers are CARNIVORES – they get their energy by eating other animals.

Respiration losses occur from each trophic level. Respiration releases the energy necessary to drive metabolic reactions, e.g. active transport and protein synthesis.

A **trophic level** is the position of an organism in a food chain or web

Food chains are short because
- long chains would need an enormous producer biomass spread over a vast area
- energy transfer between levels is so inefficient – fox might only receive 10% × 10% × 5% of light energy falling on producer.

PRODUCER   PRIMARY CONSUMER   SECONDARY CONSUMER

The Sun is the source of energy that drives all food chains. Less than 1% of the energy released from the Sun's surface actually falls onto the leaves of the photosynthesizing plants.

Decomposer organisms are fungi and bacteria that obtain their energy and raw materials from dead or waste organic material – eventually the entire energy content of these remains will be released as **heat** from inefficient respiration. Some energy may be 'stored' in fossil fuels, e.g. coal, if decomposition cannot be completed.

FOOD WEBS show a network of interconnected food chains. They are called food webs because
- each herbivore usually eats more than one type of producer
- each carnivore must feed on more than one type of herbivore.
The more complex a food web is (i.e. the more food chains involved) the more stable it is, and the less likely it is to be affected by the loss of one type of organism.

HUMANS SHOULD EAT MORE VEGETABLES!
Because of this inefficient energy transfer, it is better for humans to eat green plants and not to waste energy by feeding crop plants to animals.

Ecological pyramids

Represent numerical relationships between successive trophic levels in a community.

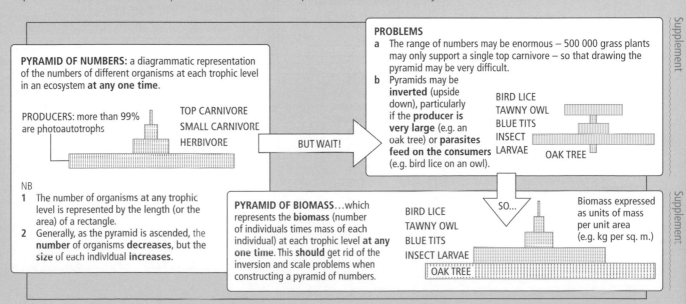

PYRAMID OF NUMBERS: a diagrammatic representation of the numbers of different organisms at each trophic level in an ecosystem **at any one time**.

PRODUCERS: more than 99% are photoautotrophs

TOP CARNIVORE
SMALL CARNIVORE
HERBIVORE

BUT WAIT!

PROBLEMS
a   The range of numbers may be enormous – 500 000 grass plants may only support a single top carnivore – so that drawing the pyramid may be very difficult.
b   Pyramids may be **inverted** (upside down), particularly if the **producer is very large** (e.g. an oak tree) or **parasites feed on the consumers** (e.g. bird lice on an owl).

BIRD LICE
TAWNY OWL
BLUE TITS
INSECT LARVAE
OAK TREE

NB
1   The number of organisms at any trophic level is represented by the length (or the area) of a rectangle.
2   Generally, as the pyramid is ascended, the **number** of organisms **decreases**, but the **size** of each individual **increases**.

SO...

PYRAMID OF BIOMASS…which represents the **biomass** (number of individuals times mass of each individual) at each trophic level **at any one time**. This **should** get rid of the inversion and scale problems when constructing a pyramid of numbers.

BIRD LICE
TAWNY OWL
BLUE TITS
INSECT LARVAE
OAK TREE

Biomass expressed as units of mass per unit area (e.g. kg per sq. m.)

Human population growth

HUMAN SUCCESS
measured as
1. worldwide distribution
2. large number of individuals
3. dominance over other species
largely due to behavioural skills which
a. allow solution of complex problems
b. allows control of / modification of environment
leading to **changes in carrying capacity of the environment.**

THREE SIGNIFICANT PHASES IN HUMAN POPULATION GROWTH

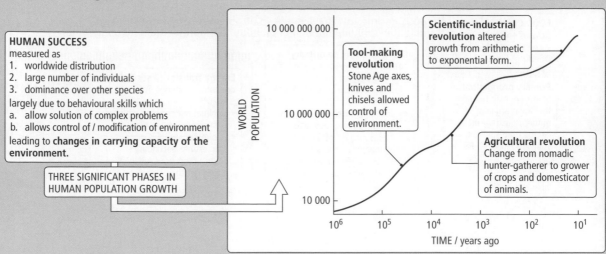

Tool-making revolution
Stone Age axes, knives and chisels allowed control of environment.

Scientific-industrial revolution altered growth from arithmetic to exponential form.

Agricultural revolution
Change from nomadic hunter-gatherer to grower of crops and domesticator of animals.

Factors affecting population growth

Birth rate and death rate 'balance' one another resulting in an equilibrium in which the population oscillates around the carrying capacity of the environment.

POPULATION EXCEEDS CARRYING CAPACITY → ENVIRONMENTAL RESISTANCE INCREASES → POPULATION FALLS

'NORMAL' POPULATION

FEEDBACK CONTROL OF POPULATION SIZE

'NORMAL' POPULATION

POPULATION LESS THAN CARRYING CAPACITY → ENVIRONMENTAL RESISTANCE DECREASES → POPULATION RISES

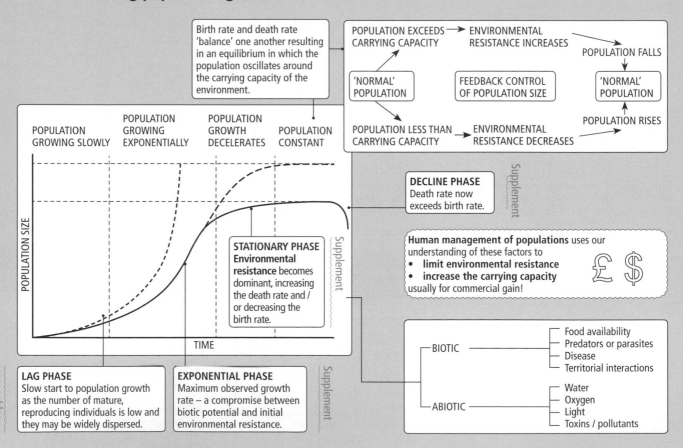

POPULATION GROWING SLOWLY

POPULATION GROWING EXPONENTIALLY

POPULATION GROWTH DECELERATES

POPULATION CONSTANT

DECLINE PHASE
Death rate now exceeds birth rate.

Supplement

STATIONARY PHASE
Environmental resistance becomes dominant, increasing the death rate and / or decreasing the birth rate.

Supplement

Human management of populations uses our understanding of these factors to
• **limit environmental resistance**
• **increase the carrying capacity**
usually for commercial gain!

£ $

LAG PHASE
Slow start to population growth as the number of mature, reproducing individuals is low and they may be widely dispersed.

EXPONENTIAL PHASE
Maximum observed growth rate – a compromise between biotic potential and initial environmental resistance.

BIOTIC
— Food availability
— Predators or parasites
— Disease
— Territorial interactions

ABIOTIC
— Water
— Oxygen
— Light
— Toxins / pollutants

Supplement

1. This diagram represents the number of different organisms in a certain food chain.

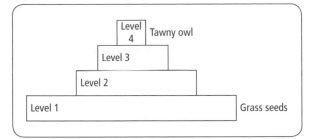

   a. What name is given to this type of diagram? (1)
   b. Study the diagram and suggest organisms that could be at levels **2** and **3**. (2)
   c. Why are there fewer organisms at level **3** than level **2**? (2)
   d. Draw out a similar diagram for the food chain:
      Oak tree ⇒ aphid ⇒ blue tit ⇒ sparrowhawk
      Explain why this diagram is not identical to the one shown above. (4)

2. Some students carried out a survey of the animals feeding on an oak tree. They collected a number of different species, and observed which animals fed on leaves and which fed on other animals. The results of their survey are shown in the table below.

ANIMAL COLLECTED / OBSERVED	NUMBER RECORDED	FOOD EATEN BY ANIMAL
Willow warbler	5	Larvae of winter moths and oak eggars
Winter moth larva	44	Oak leaves
Oak eggar caterpillar	53	Oak leaves
Tawny owl	2	Willow warblers, great tits and field mice
Great tit	5	Larvae of winter moths and oak eggars
Beetle	4	Larvae of winter moths and oak eggars
Field mouse	3	Acorns

Copy the following table. Use the information above to name the animals in the trophic (feeding) levels in this table. (2)

TROPHIC LEVEL	ORGANISMS PRESENT
4: tertiary consumers	
3: secondary consumers	
2: primary consumers	
1: producer	

   a. Calculate how many animals there were at each trophic level. Draw a table of your results. (2)
   b. Use a piece of graph paper to draw an accurate pyramid of numbers for the animals at the three trophic levels. (4)

   c. Why is the number of tertiary consumers not really representative of this particular oak tree ecosystem? (2)

3. a. A small population of rabbits was introduced to an island where rabbits had never lived before. The figure below shows the change in the size of the rabbit population over a few years.

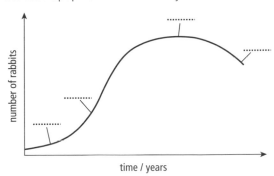

   Complete the figure by labelling the four phases of this population growth.
   • death   (use letter **D**)
   • exponential (log)   (use letter **E**)
   • lag   (use letter **L**)
   • stationary   (use letter **S**)
   Write the letters **D**, **E**, **L** and **S** on the figure in the spaces provided. (3)

   b. State three factors that could affect the rate of growth of this rabbit population. (3)

   Cambridge IGCSE Biology 0610 Paper 21 Q7 June 2012

4. The diagram shows the flow of energy through a food chain. The figures are in kilojoules of energy per square metre per year.

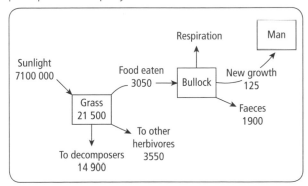

   a. How do decomposers break down carbon-containing compounds from the dead remains of grass plants? (3)

b. When the bullock eats the grass, much of the energy from the grass is released in respiration.
    i.   How much energy is released by the bullock in respiration?
         .... kJ per m² per year (1)
    ii. Give **one** use of the energy released in respiration. (1)
c. Intensive rearing of cattle indoors is an attempt to reduce energy losses. The table shows the energy balance for indoor and outdoor meat production from cattle.

	kJ PER m² PER YEAR	
	INDOORS	OUTDOORS
Energy input as food	10 000	5 950
Energy input as fossil fuel	6 000	50
Energy trapped in meat	40	1.8

    i.   The percentage efficiency of rearing cattle indoors is 0.25%. Use the following formula to calculate the percentage efficiency of rearing cattle outdoors.

$$\text{Percentage efficiency} = \frac{\text{Energy trapped in meat}}{\text{Total energy input}} \times 100$$

         Show clearly how you work out your answer.
         ...... percentage efficiency (2)
    ii. Suggest **two** reasons why rearing cattle indoors is more efficient than rearing them outdoors. (2)
    iii. Suggest **two** possible disadvantages of rearing cattle indoors. (2)

5. a. A group of pupils were studying a forest. They noticed that the plants grew in two main layers. They called these the tree layer and the ground layer.

Tree layer
Ground layer

The pupils measured the amount of sunlight reaching each layer at different times in the year. Their results are shown on the graph.

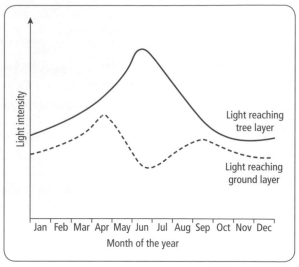

    i.   During which month did most light reach the tree layer? (1)
    ii. During which month did most light reach the ground layer? (1)
    iii. Suggest why the amount of sunlight reaching the ground layer is lower in mid-summer than in the spring. (1)
b. The pupils found bluebells growing in the ground layer. Bluebells grow rapidly from bulbs. They flower in April and by June their leaves have died.
    i.   Suggest why bluebells grow rapidly in April. (1)
    ii. Suggest why the bluebell leaves have died by June. (1)
c. The pupils hung buckets under the trees and collected the petals and fruit that fell off the trees during the year.
    The results are shown on the bar chart.

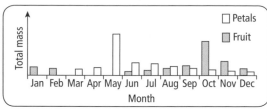

    i.   The trees reproduced by sexual reproduction. Suggest in which month most flowers are fertilised. (1)
    ii. The pupils worked out that it took an average of five months for a fruit to grow and be dispersed. Suggest how this could be worked out from their bar chart. (2)

ECOLOGY AND ECOSYSTEMS: Crossword

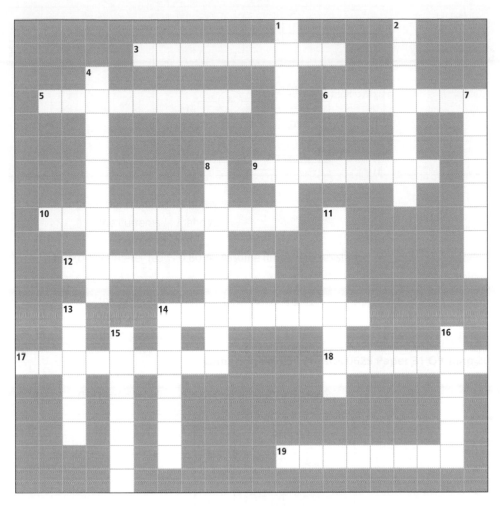

ACROSS:
3 A unit of energy
5 The living organisms and their non-living environment
6 Piece of apparatus that can separate a piece of the environment for sampling
9 An organism that feeds on the molecules made by producers
10 An organism that must obtain its food molecules ready-made
12 All the living organisms in a habitat
14 A straight-line way of showing a series of feeding relationships
17 An organism that feeds only on plants
18 A part of the environment that can provide food, shelter and breeding sites
19 An organism that feeds on both plant and animal material

DOWN:
1 An organism that must make its own organic compounds from simple raw materials
2 The stage in a food chain that can convert light energy into chemical energy
4 All the organisms of the same species living in a particular area
7 A survey method that is useful for showing changes in living organisms when moving from one habitat to another
8 A meat eater
11 The source of energy for every food chain
13 This element is cycled during respiration and photosynthesis
14 A way of showing interlinked and overlapping food chains
15 Is calculated from number of organisms times mass of each individual
16 A representative part of a population

The carbon cycle

involves **interconversion of simple and complex molecules**

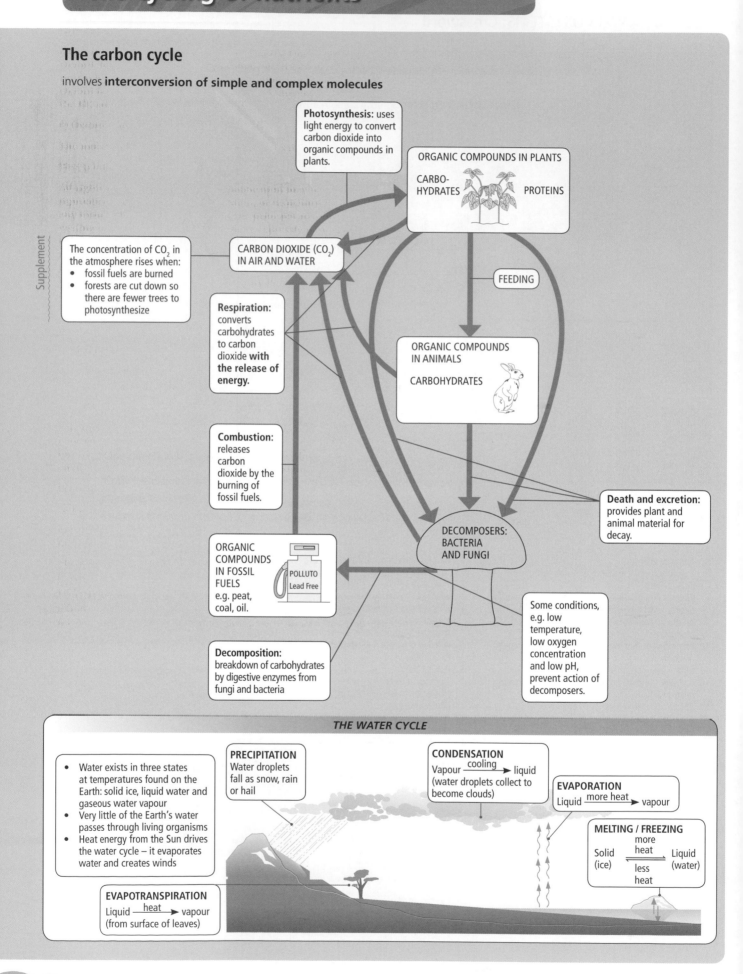

Photosynthesis: uses light energy to convert carbon dioxide into organic compounds in plants.

ORGANIC COMPOUNDS IN PLANTS
CARBO-HYDRATES   PROTEINS

The concentration of CO_2 in the atmosphere rises when:
• fossil fuels are burned
• forests are cut down so there are fewer trees to photosynthesize

CARBON DIOXIDE (CO_2) IN AIR AND WATER

FEEDING

Respiration: converts carbohydrates to carbon dioxide **with the release of energy.**

ORGANIC COMPOUNDS IN ANIMALS

CARBOHYDRATES

Combustion: releases carbon dioxide by the burning of fossil fuels.

Death and excretion: provides plant and animal material for decay.

ORGANIC COMPOUNDS IN FOSSIL FUELS e.g. peat, coal, oil.

POLLUTO Lead Free

DECOMPOSERS: BACTERIA AND FUNGI

Some conditions, e.g. low temperature, low oxygen concentration and low pH, prevent action of decomposers.

Decomposition: breakdown of carbohydrates by digestive enzymes from fungi and bacteria

THE WATER CYCLE

• Water exists in three states at temperatures found on the Earth: solid ice, liquid water and gaseous water vapour
• Very little of the Earth's water passes through living organisms
• Heat energy from the Sun drives the water cycle – it evaporates water and creates winds

PRECIPITATION
Water droplets fall as snow, rain or hail

CONDENSATION
Vapour $\xrightarrow{\text{cooling}}$ liquid (water droplets collect to become clouds)

EVAPORATION
Liquid $\xrightarrow{\text{more heat}}$ vapour

MELTING / FREEZING
Solid (ice) $\underset{\text{less heat}}{\overset{\text{more heat}}{\rightleftarrows}}$ Liquid (water)

EVAPOTRANSPIRATION
Liquid $\xrightarrow{\text{heat}}$ vapour (from surface of leaves)

The nitrogen cycle also

involves **interconversion of simple and complex molecules**

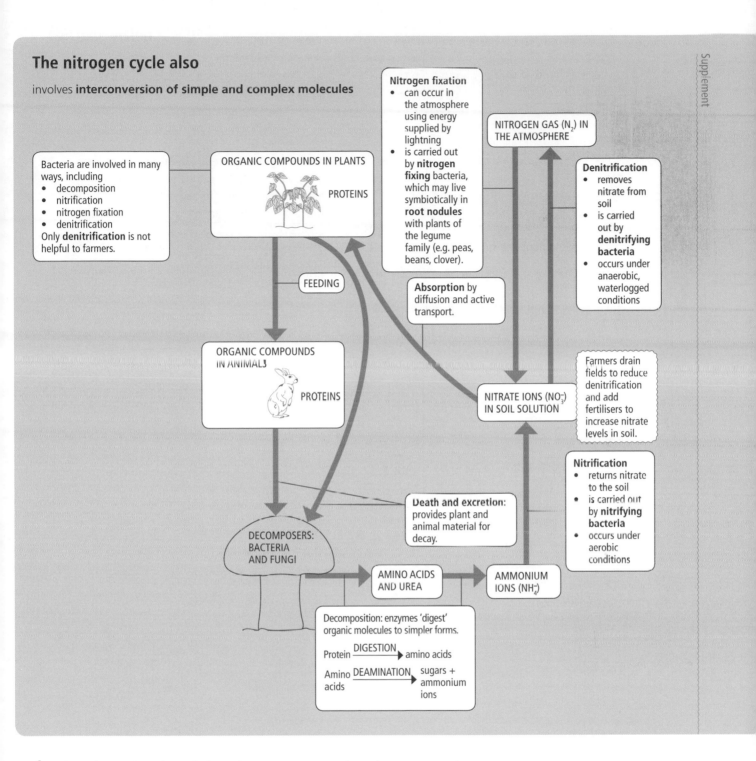

Bacteria are involved in many ways, including
- decomposition
- nitrification
- nitrogen fixation
- denitrification

Only **denitrification** is not helpful to farmers.

ORGANIC COMPOUNDS IN PLANTS

PROTEINS

Nitrogen fixation
- can occur in the atmosphere using energy supplied by lightning
- is carried out by **nitrogen fixing** bacteria, which may live symbiotically in **root nodules** with plants of the legume family (e.g. peas, beans, clover).

NITROGEN GAS (N$_2$) IN THE ATMOSPHERE

Denitrification
- removes nitrate from soil
- is carried out by **denitrifying bacteria**
- occurs under anaerobic, waterlogged conditions

FEEDING

Absorption by diffusion and active transport.

ORGANIC COMPOUNDS IN ANIMALS

PROTEINS

NITRATE IONS (NO$_3^-$) IN SOIL SOLUTION

Farmers drain fields to reduce denitrification and add fertilisers to increase nitrate levels in soil.

Nitrification
- returns nitrate to the soil
- is carried out by **nitrifying bacteria**
- occurs under aerobic conditions

DECOMPOSERS: BACTERIA AND FUNGI

Death and excretion: provides plant and animal material for decay.

AMINO ACIDS AND UREA

AMMONIUM IONS (NH$_4^-$)

Decomposition: enzymes 'digest' organic molecules to simpler forms.

Protein $\xrightarrow{\text{DIGESTION}}$ amino acids

Amino acids $\xrightarrow{\text{DEAMINATION}}$ sugars + ammonium ions

---

1. The carbon cycle is shown below. The arrows represent the various processes that happen in the cycle.

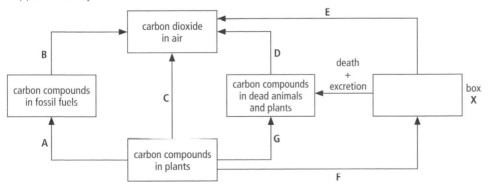

a. i.   Copy and complete the diagram by filling in
         box **X**.   (1)

    ii   State the letters of **two** arrows that represent
         respiration.   (2)

    iii State the letter of the arrow that can only
         represent combustion in this cycle.   (1)

    iv   State the letter of the arrow that represents the
         process in the cycle that takes millions of years
         to happen.   (1)

b. i.   Photosynthesis is not shown on the diagram.
         Draw an arrow to represent photosynthesis and
         label it **P**.   (1)

    ii   Write a word equation for photosynthesis.   (2)

    Cambridge IGCSE Biology 0610 Paper 2 Q5 June 2007

2. The diagram below shows the feeding process in a small
    organism.

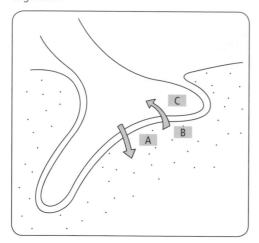

The organism is feeding on some waste protein – the
skin of a dead animal.

a.   Which **class** of enzyme must be secreted at **A**?   (1)
b.   Which type of compound will be absorbed at **B**? (1)
c.   Which Kingdom does this organism belong to?   (1)
d.   What is the name given to the feeding structure
     labelled **C**?   (1)

A biologist was interested in the uptake of the
compounds at **B**. He carried out an experiment to
investigate the effect of oxygen concentration on this
process. The results are shown in the table below.

OXYGEN CONCENTRATION / %	0	4	8	12	16	20
RATE OF UPTAKE / ARBITRARY UNITS	2	10	18	25	32	32

e.   Draw a graph of this information.   (5)
f.   What does the graph tell you about the uptake of
     compounds at **B**?   (3)
g.   Suggest another factor that might affect the rate
     of uptake of these compounds.   (1)
h.   How could this information be useful to an organic
     farmer?   (2)

3. This diagram shows part of the nitrogen cycle.

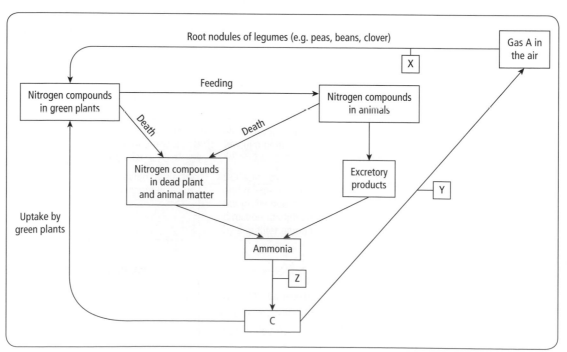

a. Identify the gas labelled **A**. (1)
b. Many of the stages of the nitrogen cycle depend on the actions of bacteria. Three of these processes are nitrification, nitrogen fixation and denitrification. Match up the labels **X**, **Y** and **Z** with these three processes. (3)
c. Identify the compound labelled **C**. (1)
d. Name one biological molecule that is made by plants from compound **C**. (1)
e. Animal excretory products are broken down in the nitrogen cycle. Name **one** animal excretory product containing nitrogen. (1)

4. a. The figure below shows a water cycle.

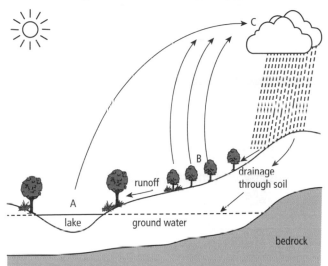

i. Name the processes happening at **A**, **B** and **C**. (3)
ii. Suggest why the most rainfall occurs over hills and mountains. (1)
b. Lakes are often naturally rich in nutrients such as nitrates.
Using information from the figure, suggest how these nutrients are moved from the hill into the lake. (1)

Cambridge IGCSE Biology 0610 Paper 21 Q4 November 2011

5. The figure below shows a carbon cycle.

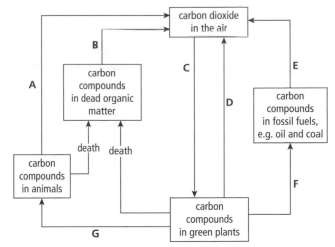

a. i. Name the process represented by arrow **A**. (1)
   ii. Name the process represented by arrow **E**. (1)
b. i. Name **one** group of organisms responsible for process **B**. (1)
   ii. List **two** environmental conditions needed for process **B** to occur. (1)
c. i. Which arrow represents photosynthesis? (1)
   ii. Complete the word equation for photosynthesis.

   .... + ... → oxygen + ... (2)

   iii. This process needs a supply of energy. Name the form of energy needed. (1)

Cambridge IGCSE Biology 0610 Paper 21 Q5 November 2012

CYCLING OF NUTRIENTS: Crossword

ACROSS:

1 Loss of nitrogen from nitrogen compounds to the atmosphere
3 The process that converts 4 across to carbohydrates
4 The most common carbon-containing gas (6,7)
5 Conversion of ammonium salts to nitrate
8 Process that makes nitrogen gas available as nitrogen-containing compounds
9 Underground homes of bacteria responsible for 8 across
12 Releases 4 across as foods are oxidized
14 Decomposers that have bodies made of hyphae
15 Microbes responsible for much of the cycling of nutrients
16 An organism that breaks down complex molecules into simpler ones
17 One way of returning animal wastes to the environment
18 This type of transport is needed for the uptake of many minerals
19 Simple nitrogen-containing compound in living organisms (5,1)

DOWN:

2 The over-feeding of lakes and ponds with excess nitrates
6 Molecules needed by 16 across to break down large molecules
7 The washing of nitrates from the soil into water
10 Complex organic compounds: contain nitrogen and are needed for growth
11 Process that releases 4 across from fossil fuels
13 Important ionic form of nitrogen

Chapter 34:
Useful microbes

Microbes (fungi, bacteria, viruses and moulds) can be useful

SEWAGE TREATMENT

Decay bacteria carry out AEROBIC DIGESTION of solutes dissolved in water

COMPLEX COMPOUNDS $\rightarrow$ CO_2, H_2O, H_2S and ammonium compounds

nitrifying $\downarrow$ bacteria

NITRATES

ANAEROBIC DIGESTION

COMPLEX COMPOUNDS $\rightarrow$ Fatty acids, amino acids and sugars

Also produce METHANE (in BIOGAS), an important fuel

FUELS

- GASOHOL – yeast produces alcohol from sugar cane waste. The alcohol is mixed with gasoline to produce gasohol, a fuel for cars and lorries
- BIOGAS – bacteria ferment plant and animal waste to produce a mixture of methane and carbon dioxide. This biogas can be burned to release energy e.g. in simple home cookers

SOURCE OF ENZYMES

- LIPASE – used to digest fatty / greasy stains when part of a BIOLOGICAL WASHING POWDER
- PECTINASE – breaks down 'lumps' of fruit to clear fruit juices
- LACTASE – used to digest lactose to produce lactose-free milk

VACCINES

- Some are produced in genetically modified yeast
- 'WEAKENED' bacteria can be used as vaccines themselves

ANTIBIOTICS

Moulds (e.g. *Penicillium*) can be grown in bioreactors to produce compounds (e.g. penicillin) that can control the reproduction of harmful bacteria

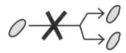

GENETIC ENGINEERING

Bacteria can be genetically modified to produce proteins (e.g. insulin) which are useful to humans. Bacteria are useful because:
- they have a rapid rate of reproduction
- they can make complex organic molecules

VIRUSES can carry helpful genes to repair damage: GENE THERAPY

BREWING AND BAKING

Yeast

GLUCOSE $\longrightarrow$ ALCOHOL + CARBON DIOXIDE

Anaerobic conditions

is poisonous, and eventually kills the yeast!

'fizz' in some drinks (e.g. beer)

'bubbles' that make bread rise

Genetic engineering (recombinant DNA technology)

- depends on enzymes and culture of microorganisms
- changes the genetic material of an organism by removing, changing or inserting individual genes.

Useful genes can also be inserted into plants
- to give resistance to insect pests
- to give resistance to herbicides
- to produce crop plants with additional vitamins

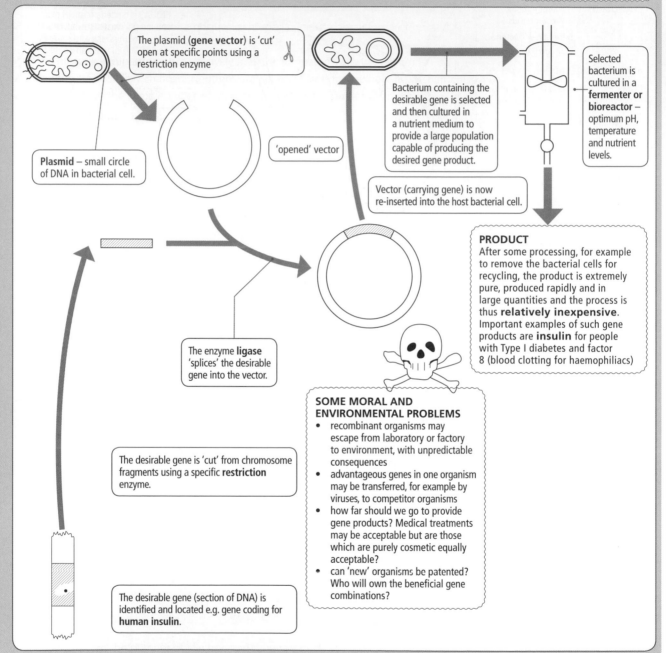

The plasmid (**gene vector**) is 'cut' open at specific points using a restriction enzyme

Plasmid – small circle of DNA in bacterial cell.

'opened' vector

Bacterium containing the desirable gene is selected and then cultured in a nutrient medium to provide a large population capable of producing the desired gene product.

Selected bacterium is cultured in a **fermenter or bioreactor** – optimum pH, temperature and nutrient levels.

Vector (carrying gene) is now re-inserted into the host bacterial cell.

The enzyme **ligase** 'splices' the desirable gene into the vector.

The desirable gene is 'cut' from chromosome fragments using a specific **restriction** enzyme.

The desirable gene (section of DNA) is identified and located e.g. gene coding for **human insulin**.

PRODUCT
After some processing, for example to remove the bacterial cells for recycling, the product is extremely pure, produced rapidly and in large quantities and the process is thus **relatively inexpensive**. Important examples of such gene products are **insulin** for people with Type I diabetes and factor 8 (blood clotting for haemophiliacs)

SOME MORAL AND ENVIRONMENTAL PROBLEMS
- recombinant organisms may escape from laboratory or factory to environment, with unpredictable consequences
- advantageous genes in one organism may be transferred, for example by viruses, to competitor organisms
- how far should we go to provide gene products? Medical treatments may be acceptable but are those which are purely cosmetic equally acceptable?
- can 'new' organisms be patented? Who will own the beneficial gene combinations?

1. The industrial production of many proteins is made easier by the use of bioreactors such as the one shown in the diagram below.

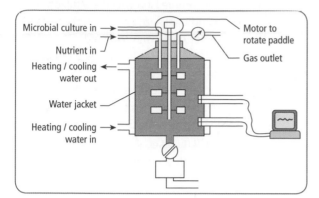

Microbial culture in →
Nutrient in →
Heating / cooling water out ←
Water jacket
Heating / cooling water in →
Motor to rotate paddle
Gas outlet

a. Why is it important that the water jacket keeps the temperature close to 30°C? (2)
b. What is the function of the paddles? (1)
c. An experiment was carried out to find the best conditions for producing the protein. Samples of the liquid inside the fermenter were taken at five-hour intervals, and analysed for the protein product. The results of the experiment are shown in the table below.

TIME / HOURS	MASS OF PRODUCT / KG
5	2.12
10	2.50
15	2.80
20	3.04
25	3.28
30	3.40
35	3.50
40	3.50
45	3.50

i. Plot these results in the form of a graph. (4)
ii. The company collects the protein as soon as the yield reaches 90 per cent of its maximum. At what time can the product be collected? Show your working. (3)

2. Antibiotics are useful drugs. The antibiotic, amoxycillin, can be manufactured by growing a mould in a nutrient solution in a fermenter.
The graph shows how the concentration of the nutrient changes over time, in a fermenter.

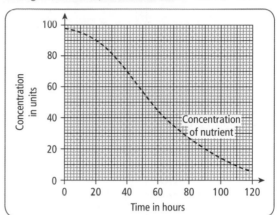

a. The table shows how the concentration of amoxycillin changes in the fermenter.

TIME IN HOURS	0	20	40	60	80	100	120
CONCENTRATION OF AMOXYCILLIN IN UNITS	0	1	57	86	93	98	99

On the grid above, draw the graph for amoxycillin production. (2)
b. Explain why the nutrient concentration in the fermenter changes over time. (1)
c. Describe the relationship between the concentration of nutrient and the concentration of amoxycillin (2)
d. Why do doctors give their patients antibiotics? (1)

3. The diagram shows a fermenter that may be used to commercially produce an antibiotic. The culture medium is infected with an aerobic antibiotic-producing mould.

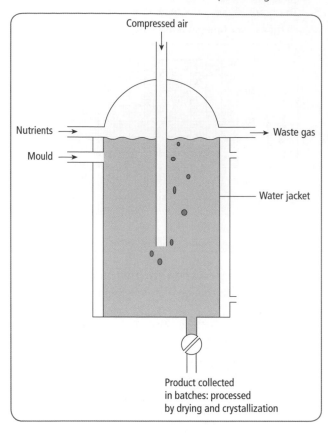

Compressed air

Nutrients →

Mould →

Waste gas

Water jacket

Product collected in batches: processed by drying and crystallization

a. i. What is meant by the term *aerobic*? (1)

   ii. Why is it necessary to supply the mould with glucose? (1)

   iii. Suggest **one** reason why the antibiotic is dried and crystallized once the contents of the fermenter have been collected. (1)

b. Name **one** antibiotic that could be produced in this system. (1)

c. Sore throats are sometimes caused by a bacterium called *Streptococcus*. This type of sore throat can be treated with an antibiotic, but there is a danger that the bacterium can develop a resistance to the antibiotic. The diagram shows how an antibiotic-resistant form may develop.

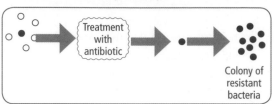

Treatment with antibiotic

Colony of resistant bacteria

   i. Name the process that leads to the **formation** of the antibiotic-resistant bacterium. (1)

   ii. Name the process that leads to the development of a colony of antibiotic-resistant bacteria while killing off the non-resistant bacterial cells. (1)

d. Explain the difference between an antiseptic and an antibiotic. (2)

4. The diagram shows how human factor 8 can be made by genetic engineering.

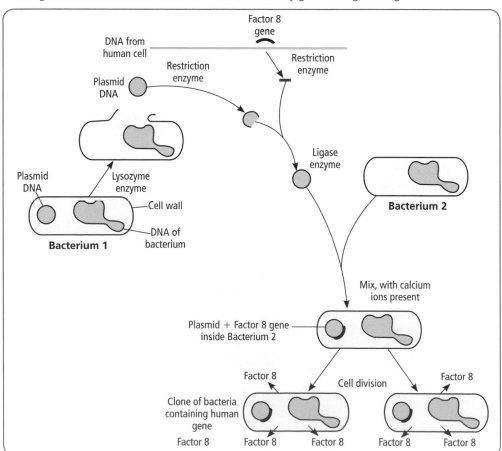

Factor 8 gene

DNA from human cell

Restriction enzyme

Restriction enzyme

Plasmid DNA

Ligase enzyme

Plasmid DNA

Lysozyme enzyme

Cell wall

DNA of bacterium

Bacterium 1

Bacterium 2

Mix, with calcium ions present

Plasmid + Factor 8 gene inside Bacterium 2

Factor 8

Cell division

Factor 8

Clone of bacteria containing human gene

Factor 8   Factor 8   Factor 8   Factor 8   Factor 8

a. Using only the information from the diagram describe how the gene for factor 8 is put into a bacterium. (7)

b. Why is factor 8 valuable? (1)

c. Name one other human protein that can be made by genetic engineering, and state why it is so useful. (2)

d. A bacterium that has been altered in this way is a genetically modified organism (GMO). Why are some people against the use of GMOs? (2)

USEFUL MICROBES: Crossword

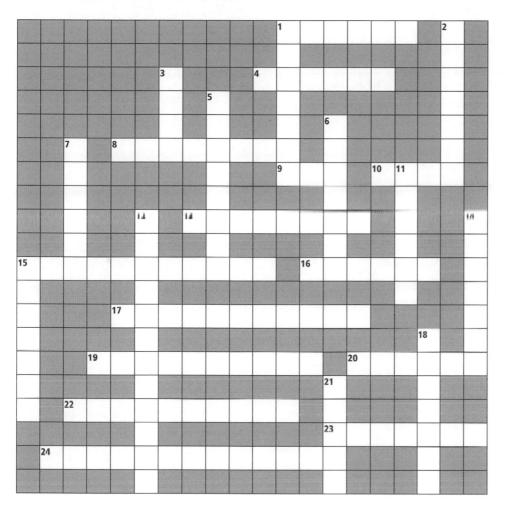

ACROSS:
1 A useful fuel released when waste materials decompose
4 A mixture of faeces, water and bacteria: must be treated before the water is safe to drink
8 Enzyme, produced by genetic engineering, that is used in cheese production
9 Abbreviation for a genetically modified organism
10 With 7 down: responsible for the acidic taste of yoghurt
13 Most widely used organisms in biotechnology, can act as 'factories' to produce proteins using human genes as instructions
15 Dried hyphae of a fungus – a healthy alternative to meat?
16 A means of carrying a gene from one organism to another
17 Enzyme used in genetic engineering to cut into DNA molecules
19 Important drug, produced by moulds, that may kill bacteria or slow down their rate of multiplication
20 Component of biological washing powders – an enzyme that helps to remove greasy stains
22 The tasty component of vinegar, produced by fermentation with *Acetobacter*
23 Toxic product of anaerobic respiration in 21 down
24 Bacterium that ferments the sugar in milk

DOWN:
1 Industry that produces 23 across
2 A small piece of bacterial DNA that can carry a gene into a bacterial cell
3 The liquid part of soured milk, removed during production of cheese
5 Important enzyme in the leather-softening process
6 With 11 down – the gaseous product of anaerobic respiration (puts the 'fizz' into many drinks)
7 See 10 across
11 See 6 down
12 Set of reactions carried out by microbes with many products useful to humans
14 Enzyme used to break down starch in the early stages of 1 across
15 Can be a useful product for powering machinery at landfill sites (although it is also a greenhouse gas)
18 A useful addition to petrol in internal combustion engines
21 Type of microbe responsible for alcoholic fermentation

REVISION SUMMARY: MATCHING TERMS

Match the words from the list below with the definitions in the table.

WHEY, FERMENTATION, YEAST, LACTOBACILLUS, CARBON DIOXIDE, ALCOHOL, MYCOPROTEIN, BACTERIUM, CHYMOSIN, ACETOBACTER, BIOREACTOR, PASTEURIZATION, STERILIZATION, DISTILLATION, HOPS

DEFINITION	MATCHING WORD
Increases proportion of alcohol by removal of water	
Anaerobic respiration	
Vessel in which growth of microbes can be controlled for industrial or medical reasons	
Bacterium that plays an important part in the production of cheese	
Gassy product of yeast fermentation – important in fizzy wines and beers	
Liquid by-product of cheese manufacture	
Adds flavour to beer	
The key organism in both brewing and baking	
Fermentation product that is useful in medicine but may lead to intoxication	
Milk treatment that kills all harmful microbes	
Important enzyme in the formation of solid cheeses	
Bacterium involved in the production of vinegar	
Treatment that kills all microbes in a food or medical product	
Healthy product from the bodies of the fungus *Fusarium graminearum*	
Important microbe to humans, even though it has no nucleus	

Chapter 35

Human impact on the environment

Impact of humans on the environment

We now know that human activities can have a major impact on the environment.

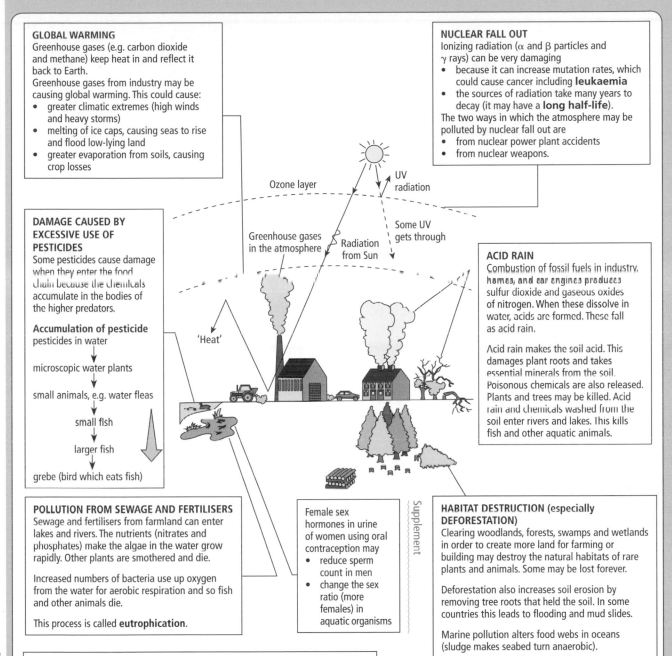

GLOBAL WARMING
Greenhouse gases (e.g. carbon dioxide and methane) keep heat in and reflect it back to Earth.
Greenhouse gases from industry may be causing global warming. This could cause:
- greater climatic extremes (high winds and heavy storms)
- melting of ice caps, causing seas to rise and flood low-lying land
- greater evaporation from soils, causing crop losses

NUCLEAR FALL OUT
Ionizing radiation (α and β particles and γ rays) can be very damaging
- because it can increase mutation rates, which could cause cancer including **leukaemia**
- the sources of radiation take many years to decay (it may have a **long half-life**).
The two ways in which the atmosphere may be polluted by nuclear fall out are
- from nuclear power plant accidents
- from nuclear weapons.

DAMAGE CAUSED BY EXCESSIVE USE OF PESTICIDES
Some pesticides cause damage when they enter the food chain because the chemicals accumulate in the bodies of the higher predators.

Accumulation of pesticide
pesticides in water
↓
microscopic water plants
↓
small animals, e.g. water fleas
↓
small fish
↓
larger fish
↓
grebe (bird which eats fish)

Ozone layer

UV radiation

Greenhouse gases in the atmosphere

Radiation from Sun

Some UV gets through

'Heat'

ACID RAIN
Combustion of fossil fuels in industry, homes, and car engines produces sulfur dioxide and gaseous oxides of nitrogen. When these dissolve in water, acids are formed. These fall as acid rain.

Acid rain makes the soil acid. This damages plant roots and takes essential minerals from the soil. Poisonous chemicals are also released. Plants and trees may be killed. Acid rain and chemicals washed from the soil enter rivers and lakes. This kills fish and other aquatic animals.

POLLUTION FROM SEWAGE AND FERTILISERS
Sewage and fertilisers from farmland can enter lakes and rivers. The nutrients (nitrates and phosphates) make the algae in the water grow rapidly. Other plants are smothered and die.

Increased numbers of bacteria use up oxygen from the water for aerobic respiration and so fish and other animals die.

This process is called **eutrophication**.

Female sex hormones in urine of women using oral contraception may
- reduce sperm count in men
- change the sex ratio (more females) in aquatic organisms

HABITAT DESTRUCTION (especially DEFORESTATION)
Clearing woodlands, forests, swamps and wetlands in order to create more land for farming or building may destroy the natural habitats of rare plants and animals. Some may be lost forever.

Deforestation also increases soil erosion by removing tree roots that held the soil. In some countries this leads to flooding and mud slides.

Marine pollution alters food webs in oceans (sludge makes seabed turn anaerobic).

NON-BIODEGRADABLE PLASTICS can pollute both aquatic and terrestrial habitats:
- block water drainage channels causing waterlogging of soil
- may be eaten by animals, and block digestive system
- release toxic particles if they are burned

1. A school party was spending some time on a rocky shore. Some of the students decided to make a survey of the different molluscs on the shore. The following diagram represents a section of the shore that had been sampled taking line transects across it.

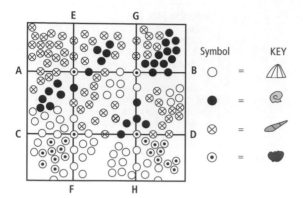

a. Which of the four transects corresponds to the following diagram? (2)

b. From the results of this transect alone, which mollusc type seems to be the most abundant? Give its symbol. (2)

c. Use the diagram and your answer to (**b**) to suggest why line transects sometimes give unreliable data. (2)

d. The section of shore shown below measures 10 metres by 10 metres. Estimate the total number of limpets on this section of shore using only the information provided by the random quadrats. (3)

Quadrats in randomly chosen sites
(each quadrat encloses one square metre)

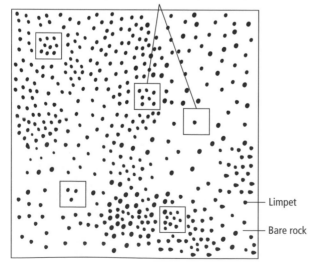

e. Oil was released from the tanks of a passing ship, and some washed up on this section of shore. The diagram below shows this section of shore four weeks after the oil spill. Using the same method, estimate the **new** total of limpets. (3)

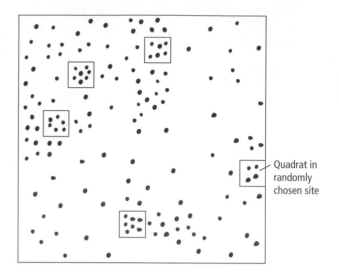

Quadrat in randomly chosen site

f. From your results, does the oil seem to have affected the limpet population? Explain your answer. (2)

g. From an overall view of the two diagrams, does the oil seem to have affected the limpet population? Suggest a source of error in the sampling technique that was responsible for failing to show up the effects of the oil. (2)

2. The diagram below shows the sequence of events associated with acid rain pollution. Copy and complete the diagram using terms chosen from the following list.

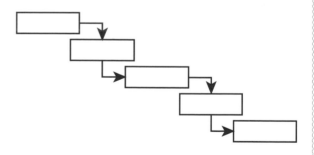

A: sulfur dioxide rises into the atmosphere
B: acid rain falls
C: fish die in acidified lakes
D: discarded car batteries leak acid
E: fuel combustion in power stations
F: ozone is produced in power stations
G: pollutant combines with water vapour

Supplement

3. Householders in the UK produce approximately 20 million tonnes of domestic waste each year. The table shows the average composition of domestic waste.

TYPE OF WASTE MATERIAL	PERCENTAGE OF TOTAL
Paper and board	32
Food and garden rubbish	24
Glass	10
Plastics	7
Metal	4
Other	

a. Complete the table, and then draw a bar graph of the information it contains. (5)

b. What is meant by the term *biodegradable*? Give **one** example of material that is biodegradable. (3)

c. Much food and garden rubbish is disposed of in plastic bags. The interior of the bag is anaerobic, and an explosive gas can be formed. What is the name of this explosive gas? (1)

d. If the 'other' category is ignored, what percentage of the total domestic waste could be biodegradable? Show your working. (3)

4. a. Draw a simple diagram to explain the meaning of the term *global warming*. (3)

b. Name **two** different greenhouse gases. (2)

c. Explain how global warming might affect
   i. sea level
   ii. the pattern of insect pest distribution around the world
   iii. the likelihood of gales and storms (6)

5. a. i. Which form of the Sun's energy is used by plants? (1)
   ii. Name the process that uses this absorbed energy. (1)

b. The graph shows how the concentration of carbon dioxide in the atmosphere has changed over a period of about 20 years.

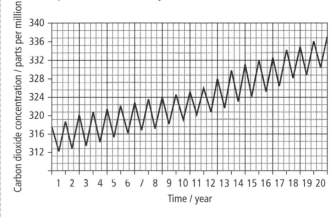

Describe the changes shown by this graph. (2)

c. The atmosphere around the earth acts as a trap for energy from the Sun.
Carbon dioxide in the air traps heat energy.
   i. Suggest the effect the overall change in the graph may be having on the Earth's climate. Explain your answer. (3)
   ii. Humans cause changes in ecosystems, including changing the amount of carbon dioxide in the atmosphere.
   Suggest **two** ways in which the overall change can be reversed. (2)

Cambridge IGCSE Biology 0610 Paper 2 Q5 November 2005

6. Deforestation occurs in many parts of the world.
a. State **two** reasons why deforestation is carried out. (2)
b. i. Explain **two** effects deforestation can have on the **carbon cycle**. (4)
   ii. Describe **two** effects deforestation can have on the **soil**. (2)
   iii. Forests are important and complex ecosystems. State **two** likely effects of deforestation on the forest ecosystem. (2)

Cambridge IGCSE Biology 0610 Paper 2 Q2 June 2006

7. a. Pollutants can affect the environment.
Draw **one** line from each pollutant listed to an effect it might have on the environment.

pollutant	effect
carbon monoxide	can cause mutations
insecticides	can cause rise of global temperature
ionising radiation	can lead to acid rain
methane	can poison top carnivores
sulfur dioxide	can reduce transport of oxygen in blood
untreated sewage	can spread diseases, such as cholera

(6)

b. Suggest **one** major source for each of the following pollutants.
   i. carbon monoxide (1)
   ii. carbon dioxide (1)
   iii. ionising radiation (1)

Cambridge IGCSE Biology 0610 Paper 21 Q2 June 2012

Supplement

REVISION SUMMARY: Fill in the gaps

Complete the following paragraphs about pollution. Use terms from the following list – you may use each term once, more than once or not at all.

PESTS, CFCs, ULTRAVIOLET LIGHT, NITRATE, BATTERIES, MESOPHYLL, CARBON DIOXIDE, ACID RAIN, INFRA-RED, METHANE, NITROGEN, PESTICIDES, CANCER, PHOTOSYNTHESIS, MUTATION, MOTOR OIL, OZONE, OXYGEN.

The greenhouse effect may cause global warming when .... radiation is trapped close to the Earth's surface by a layer of .... from ruminants, ... from aerosols and .... from the combustion of fossil fuels. Global warming may allow .... to extend their range, but may have the benefit of increasing production of food by ....

Holes in the ozone layer may result from ...., formerly used as refrigerants. These holes allow the entry of too much .... radiation, which may lead to skin .... and an increased rate of .... (especially dangerous as these may be passed on to the offspring of the affected person). The production of excess .... may also be damaging – for example, photosynthesis is reduced when the leaf .... is damaged.

Humans are also responsible for pollution of water, for example with excess .... washed out of farmland and with .... used to control the aquatic stages of pests. Atmospheric pollutants can fall as .... and so lower the pH of bodies of water – this can seriously reduce the suitability of the water for living organisms.

Pollution of land is often a result of the inefficient disposal of waste, such as ...., which contain cadmium, ..., which contains toxins that may cause skin cancer and old refrigerators, which contain .... (18)

Agriculture: Pluses and minuses

DEFORESTATION
Removal of trees
+ • provides more land for crops
  • removes habitat for pest herbivores … but …
− • removes habitat for other species
  • can affect water cycle
  • leads to rapid erosion of soil

IRRIGATION SYSTEMS
+ • provide water for growing plants, therefore removing a LIMITING FACTOR … but …
− • often withdraw water from natural sources
  • can lead to LEACHING of minerals from the soil

USE OF FERTILISERS
The availability of mineral nutrients is an important LIMITING FACTOR for crop growth. Especially
+ • NITRATE for proteins and nucleic acids: leaves
  • PHOSPHATE for nucleic acids, ATP and membranes: flowers and fruits
  • MAGNESIUM for chlorophyll: leaves and stems
So farmers ADD MORE FERTILISERS … but …
EXCESS FERTILISERS run off into ponds and streams

EUTROPHICATION

MECHANISATION
Use of tractors and harvesting machinery
+ • means cultivation can be quicker, and more land can be used … but …
− • more fossil fuels are used, and more pollution results
  Soil is compacted, so it becomes more difficult for rainwater to penetrate

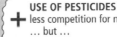

MONOCULTURE
Growing a single crop in one habitat
+ • nutrients and pest control can be exactly matched to crop
  • mechanical harvesting is easier … but ..
− • outbreaks of pests and diseases can spread very rapidly
  • loss of food chains reduces biodiversity
  • removes some nutrients specifically

MIXED CROP ROTATION
• reduces pest infestations
• improves nutrient balance
✓

USE OF PESTICIDES
+ less competition for nutrients, light and water … but …
− pesticides accumulate in food chains, with top predators being particularly affected

DDT

REMOVAL OF HEDGEROWS
+ • more growing space for crops
  • less competition for light, water and nutrients
  • fewer habitats for pests… … but …

NO NEST!
GONE

− Less shelter and fewer nesting sites for birds, pollinating insects and beetles, which are predators on pests

FAMINE (see also p. 39) may result from
• unequal distribution of food
• drought or flooding
• poverty (inability to pay for food)
• overpopulation (more competition for food)

and may cause

• disease
• mass migration of people
• war (to capture food or sites of agriculture)

Supplement

1. In order to increase the yield of crops, many farmers have used some or all of the following practices:
   a. cutting down hedges
   b. increased use of nitrate fertilisers
   c. burning of stubble after harvesting
   d. drainage of wet fields
   e. repeated growth of the same crop in the same field
   f. deep ploughing of soil using heavy machinery

Many conservationists believe that these techniques may be damaging to the environment. Choose any **two** of the above and state how they help the farmer to increase the yield of crop, then choose **three** other techniques and explain how they might damage the environment.
   Write your answers in a table like this one. (5)

TECHNIQUE	BENEFIT TO FARMER

TECHNIQUE	HARM TO THE ENVIRONMENT

2. The following diagrams represent population distributions in two different countries.

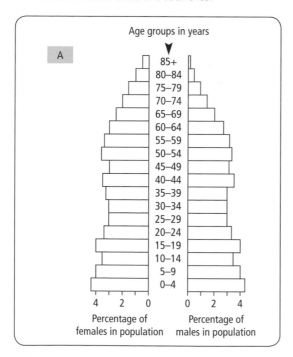

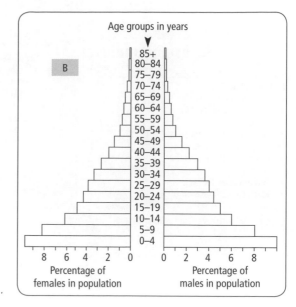

   a. In age pyramid A, what percentage of the male population are aged 60 and over? (1)
   b. In age pyramid A, which sex has the higher proportion reaching old age (65+)? (1)
   c. In age pyramid B, what proportion of the population are female aged 15-19? (1)
   d. In age pyramid B, which age group makes up exactly 4 per cent of the male population? (1)
   e. Which age pyramid represents the population of a developing country? Give **two** reasons for your answer. (2)
   f. In the country represented by age pyramid A there is a population of 60 000 000. How many are males aged 15–19? (2)
   g. Explain how the agricultural revolution allowed an increase in population of the United Kingdom. (3)

3. a. The table below shows one possible system of crop rotation used in agriculture.

FIELD	YEAR 1	YEAR 2	YEAR 3	YEAR 4
A	Sprouts	Peas	Potatoes	Fallow or cereal
B	Fallow or cereal	Sprouts	Peas	Potatoes
C	Peas	Potatoes	Fallow or cereal	Sprouts

   i. Write out the likely sequence of a fourth field, field D. (1)
   ii. What is meant by the term *crop rotation*? (2)

b. The level of nitrate in the soil can be measured. The values obtained for the sprout and the pea fields are shown in the following table.

CROP	FIELD	NITRATE LEVEL IN SOIL / ARBITRARY UNITS
Peas	A	82
Peas	B	88
Peas	C	80
Sprouts	A	54
Sprouts	B	44
Sprouts	C	52

i. Draw a bar chart of these results. (3)
ii. Calculate the percentage increase in nitrate level when peas are grown. Show your working. (3)

4. Study the revision diagram on animal husbandry and answer the following questions.
a. Why are the abiotic factors of **temperature** and **light** so important to maximizing yield? (2)
b. Suggest **three** features of the food pellets used to feed the animals. (3)
c. Why is it important to weigh the animal at regular intervals? (2)
d. How could liquid slurry have a bad effect on the environment if it was allowed to run into nearby rivers? (3)
e. What are the benefits to productivity of regular veterinary care? (2)

5. Water is essential for living organisms, but there are real worries that farmers' overuse of fertilisers is polluting our water supplies. It is estimated that in the United Kingdom only about half of the fertiliser applied to the soil is actually taken up by plants. Much of it is washed away by rain and drains through the soil.
a. State **two** ways in which water is used by living organisms. (2)

b. Which ion is most common in fertilisers added by farmers? (1)
c. The concentrations of nitrate and oxygen in a polluted river are shown in the table below.

YEAR	OXYGEN CONCENTRATION / MG DM^{-3}	NITRATE CONCENTRATION / MG DM^{-3}
1959	5.9	8.4
1964	5.4	10.7
1971	4.5	12.0
1980	4.1	12.9
1988	3.9	13.8

i. Plot a bar chart of this information (5)
ii. Calculate the percentage decrease in oxygen concentration between 1959 and 1988. Show your working. (2)

d. The nitrates and phosphates in fertilisers encourage the growth of simple plants (algae) in the surface layers of the water. A similar effect is caused by the outflow of phosphate detergents. The increased numbers of algae prevent light reaching the lower levels of the water, and rooted plants die. Their remains are broken down by bacteria, which use up large quantities of dissolved oxygen. Fish and other organisms with a high oxygen requirement also die, and are also broken down by microbes, so that the problem gets worse and worse.
i. What name is given to this 'over-feeding' of algae in rivers and lakes? (1)
ii. In unpolluted waters the algae do not grow so quickly. What do you think is limiting their growth? (1)
iii. In some developing countries the use of fertilisers on rice paddies has affected local lakes and rivers. Fish have died or have migrated to less-polluted waters. What important nutrient for humans would fish supply in a developing country? (1)

REVISION SUMMARY: Match the terms with their definitions

	TERM		DEFINITION
A	Age pyramid	A	A mineral nutrient added to increase crop yield
B	Biodiversity	B	The cultivation of a single type of crop plant
C	Crop rotation	C	Plant type able to host bacteria capable of nitrogen fixation
D	Eutrophication	D	Some resource in the environment that reduces the growth of an individual or population
E	Fertiliser	E	A way of showing how many members of a population belong in each age group
F	Legume	F	The amount of growth (biomass added) that takes place over a fixed period of time
G	Limiting factor	G	Growth of algae in ponds and rivers due to 'overfeeding' with nitrate and phosphate
H	Monoculture	H	A chemical used to control the growth and reproduction of organisms that compete for human crops
I	Pesticide	I	The range of different species of living organisms
J	Productivity	J	The regular changing of the type of food plant grown in a particular field over a period of years

Conservation of species

IMPORTANT DEFINITIONS

Sustainable resource: a resource which can be removed from the environment without it running out, e.g. wind
Non-renewable resource: a resource which cannot be replaced and so must be conserved, e.g. fossil fuels
Recycling: re-using waste products or materials which would otherwise be thrown away, e.g. paper, glass, plastic and metal
Biodegradable: a material which can be broken down (decomposed) by the action of living organisms, e.g. waste food by bacteria
Sustainable development: development which provides for the needs of a growing human population without harming the environment.
This involves:
* management of conflicting demands
* planning and co-operation at local, national and international levels

There are several possible strategies

PRESERVATION: involves keeping some part of the environment **without any change**. Only possible if the area can be fenced off / protected

RECLAMATION: involves the restoration of **damaged** habitats. Often applies to the recovery of former industrial sites, such as mine workings

CREATION: involves the production of **new** habitats. Only really possible with small areas, e.g. digging a garden pond, or planting of a new forest

DIRECT PROTECTION MAY BE NECESSARY

Even when a suitable **habitat** (able to provide food, **shelter** and **breeding sites**) is available, individual species may be at direct risk from humans.

* Rhinoceroses may be hunted for their horns, mistakenly believed to have medicinal or aphrodisiac properties
* Elephants may be hunted for their ivory tusks.
* Primates (e.g. chimpanzees) and other species may be hunted as 'bush meat'
* Butterflies, molluscs (shells) and plants may be 'collected'
* Parrots, primates and fish may be collected for the pet trade

> ALL INVOLVE MANAGEMENT OF A HABITAT

> ONE USEFUL TECHNIQUE INVOLVES…

PRESSURES ON A HABITAT

* Humans have a significant **biotic** impact on the environment

Temporary when humans were **nomadic**: the environment has periods of recovery

Permanent as humans became **cultivators** and **settlers**

— Use of tools and domestication of animals → more efficient agriculture / support for larger population
— Greater demands for shelter, agricultural land and fuel → greater rate of deforestation
— Development of fossil fuels → more use of machines / greater 'cropping' / larger populations
— Pollution as greater use of fossil fuels / pesticides / fertilisers and development of nuclear power

FLAGSHIP SPECIES

Large, attractive and 'cuddly' species attract funding from agencies and donations from the public which can offer protection for less attractive species (e.g. beetles and worms) that live in the same habitat.

MANAGEMENT IS A COMPROMISE!
* Maintenance of a particular habitat (e.g. chalk hillside for wild flowers / butterflies) often means **halting succession**.
* The requirements for wildlife must be balanced by human demands for **resources** (e.g. mining for uranium), **recreation** (e.g. diving around coral reefs) and **agricultural** land.

1.

DDT AND HERONS IN UGANDA

Nine clutches of eggs from the green heron were collected close to the Kasinga Channel in Southern Uganda, during 2002. The sites were close to sample sites checked in 1990. Analysis of the eggs showed that they had an average increase in DDT concentration of 60 per cent, and the eggshells were nearly 30 per cent thinner compared with 1990. The lilac breasted roller, which lives in drier, cattle-rearing areas, has seriously declined in numbers over the same period: this bird feeds on insects taken from tree trunks in areas where DDT is used to control tsetse flies.

a. What was the average annual increase in DDT concentration in the Green Heron eggs? Show your working. (2)

b. Why is it difficult to estimate the effects on the whole population from the data available? (1)

c. Explain why high levels of DDT are thought to reduce breeding success of these birds. (2)

d. Explain why the Green Heron is likely to take up high levels of DDT. (2)

e. Suggest how the feeding method of the roller make it particularly at risk from DDT. (1)

f. In the United Kingdom, DDT use was blamed for a 60 per cent reduction in the number of sparrowhawks in the 1960s. By 1990 sparrowhawk numbers had almost recovered. Explain why sparrowhawk numbers recovered in the UK, but Green Heron numbers in Uganda are less likely to recover. (2)

2. Toads are amphibians. Only two species are native to Britain, the Common toad (*Bufo bufo*) and the Natterjack toad (*Bufo calamita*).

Natterjack toads like warm sandy soil in open and sunny habitats, with shallow pools for breeding. Examples of these habitats are heathland and sand dunes. Common toads like cooler, more shady habitats, such as woodland.

Many areas of sand dunes are being developed for camp sites. Heathland can easily change to woodland as trees grow on it. In the summer, woodland is colder than heathland due to the shade the trees create.

These conditions suit the Common toad, but not the Natterjack. As a result of the changing habitats the Natterjack toad is becoming an endangered species.

a. i. Name **one** external feature that identifies an animal as an amphibian. (1)

ii. Amphibians are a class of vertebrate. Name **two** other vertebrate classes. (2)

b. State **one** piece of information from the passage to show that the Common toad and Natterjack toad are closely related species. (1)

c. From the information provided, state **two** reasons why Natterjack toads are becoming endangered. (2)

d. Suggest measures that could be taken to protect the Natterjack toad from extinction. (2)

Below is a food web for British toads.

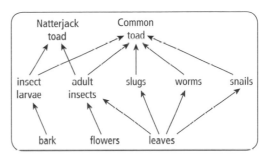

e. i. State the trophic level of toads. (1)

ii. State which foods the two species of toad both eat. (1)

iii. With reference **only** to food, suggest why the Common toad is more likely to survive when the two species are in competition. (1)

Cambridge IGCSE Biology 0610 Paper 3 Q1 November 2005

3. The Ruddy duck, *Oxyura jamaicensis*, is a native of America. A flock of 20 birds was introduced into Britain from America before 1950. The original flock settled quickly in their new habitat and started breeding. Numbers now exceed 6000.

The White-headed duck, *Oxyura leucocephala*, (a native of Spain) is a closely related species to the Ruddy duck. Female White-headed ducks are more attracted to mate Ruddy ducks than to males of their own species. Cross-breeding between the two species produces a new variety of fertile duck.

The White-headed duck is now threatened with extinction. Some conservationists are considering a plan to kill the British population of Ruddy ducks to prevent the White-headed duck becoming extinct.

The diagram below shows a male Ruddy duck.

a. State **two** features, visible in the diagram, that distinguish birds, such as the Ruddy duck, from other vertebrate groups. (2)

b. i. With reference to an example from the passage, describe what is meant by the term *binomial system*. (2)

   ii. State **two** reasons, based on information in the passage, why the Ruddy duck and White-headed duck are considered to be closely related. (2)

c. i. Explain why Ruddy ducks would **not** become extinct, even if British conservationists carried out their plan. (1)

   ii. Suggest **one** factor, other than the breeding habits of the Ruddy duck, that could result in the extinction of a bird such as the White-headed duck. (1)

d. The Ruddy duck feeds on seeds and insect larvae. The ducks are eaten by foxes and humans. Explain why these feeding relationships can be displayed in a food web, but not in a food chain. (2)

Cambridge IGCSE Biology 0610 Paper 3 Q6 June 2005

4. GOLDEN LION TAMARINS AND ZOOS

Golden lion tamarins are small South American primates, living in the coastal forests of Brazil. They require extensive forests of tall trees as they have to range widely to find food, and must avoid predators both on the ground and from the air. Many of these forests have been cut down for timber and for cattle ranching.

They breed slowly and have small litters. The mortality rate among the young animals is high, mainly because of predation.

Golden lion tamarins have been kept in zoos for a number of years but there are problems with the captive breeding programme: many of the offspring are genetically related to one another, causing a reduction in breeding rate, and adults cannot learn the tree-climbing and feeding skills they need when confined to zoos.

These animals are very attractive and suffer from poaching for their skins, even though trade in these skins is banned by CITES.

a. Give **two** features of the Tamarin that would confirm that it is a mammal. (2)

b. Why is the Golden Lion Tamarin in danger of becoming extinct? (4)

c. Give **three** arguments against keeping Golden Lion Tamarins in zoos. (3)

d. Give **two** arguments in favour of keeping Golden Lion Tamarins in zoos. (2)

e. What part does the organization CITES play in the conservation of species? (2)

f. The Golden Lion Tamarin is a 'flagship species'. What does this mean? (2)

5. Read the following article, and then answer the questions that follow it.

Following the Second World War the British government made plans to increase crop yields in Britain. Between 1945 and 1965 hedgerows were being removed at a rate of 3000 miles per year – this rate of removal increased to 5000 miles per year by 1968. The rate of removal has fallen recently, and a recent report suggests that between 1980 and 1990, 10 000 miles were removed but 5000 miles were replanted. Conservationists argue that new hedgerows are a much poorer habitat for wildlife than older, mature hedges.

In Cambridgeshire, farmers estimated that the removal of one mile of hedgerow provided an extra three acres of arable land and reduced by 40 per cent the time taken to harvest a field of cereals.

The British government spends over £2 billion per year to buy and store surplus food produced in the UK. Much of the food is eventually destroyed, or given away at very low prices. The European Union is now encouraging farmers to produce less food.

a. What was the highest rate of hedge removal before 1969? (1)

b. Why is it less important now to gain extra arable land than it was immediately after the Second World War? (2)

c. English Nature, an important conservation organization, publishes a list of important points about hedgerows. These are some of the points included in the list:

i. hedges obstruct the efficient use of modern farm machinery

ii. hedges are an attractive feature of the rural landscape

iii. hedges provide cover for game birds such as pheasant, partridge and quail

iv. hedges can provide a habitat for weeds, rabbits and insect pests

v. hedges are very expensive to maintain, barbed wire is very cheap to maintain

vi. hedges shade part of the crop, and may compete for water

vii. hedges prevent wind damage to crops

viii. hedges provide an important habitat for insects that can help with pollination and with biological control of pests

Choose **three** statements from this list that would support hedge removal, and **three** points that could be used to support arguments against hedge removal.

Three statements that support hedge removal are ....

Three statements that argue against hedge removal are .... (6)

CONSERVATION: Crossword

ACROSS:

1 The use, by humans, of a habitat for leisure purposes

4 Symbol of the Worldwide Fund for Nature – an important example of 7 down

6 Interlinked set of feeding relationships – easily disturbed if one species is removed (4,3)

8 Returning a damaged habitat to its original purpose – an important type of conservation

9 Abbreviation for the international organization that bans the trade in endangered species

13 A type of conservation in which new habitats are produced

14 A part of the environment that can provide food, shelter and breeding sites for living organisms

16 With 2 down – one method for increasing the population of an endangered species

17 Large member of the cat family that is endangered because some forms of medicine believe its body parts can treat a range of diseases

18 Humans may damage a habitat as they seek these

19 A place where animals may be protected, observed and where 2 down may take place

20 Large herbivore hunted for the ivory trade

DOWN:

2 See 16 across

3 With 7 down – an important organism in conservation as it has enough public appeal to help raise funds

5 As humans became more efficient at this, more land was taken away from wildlife

7 See 3 down

10 Large herbivore, endangered because some people believe that its horn has aphrodisiac properties

11 All conservation involves this – this is how humans, especially scientists, can be useful in conservation

12 Type of animals collected for its beautiful shell

15 Powerful insecticide that can become concentrated in a food chain and reduce the breeding success of top carnivores

IGCSE Biology study guide answers

Cambridge International Examinations bears no responsibility for the example answers to test-style questions which are contained in this publication. The example answers, marks awarded and comments that appear in this book were written by the author. In examination, the way marks would be awarded to answers like these may be different.

CHAPTER	QUESTION	ANSWER(S)
6: THE VARIETY OF LIFE	1	arachnid; mollusc; bird; mammal; insect
	2	a. i. MAMMAL; ii. FUNGUS iii. REPTILE; iv. INSECT; v. BACTERIUM; vi. VIRUS; vii. BIRD b. i. respiration; ii. excretion; iii. sensitivity / irritability
	3	a. i. B – parallel veins on leaf; ii. D – worm with body divided into segments; iii. E – body in two parts / four pairs of legs; iv. G – armoured exoskeleton / more than four pairs of limbs / limbs very specialized b. Magnification = measured / actual length so magnification = 50 / 5 = × 10
	4	a. i. antennae / three parts to body (head, thorax and abdomen) / three pairs of legs ii. gills (extract oxygen from water) b. A = Ecdyonurus; B = Potomanthus; C = Ephemera; D = Paraleptophlebia; E = Centroptilum
	5	a. C is the smallest; J is the largest b. A, D, F, G and I (all arthropods) c. i. A, G and I ii. wings / three pairs of legs
	6	a. include 'body not covered by a shell' as second alternative to Q.2 and add 'go to 4'; Q.3 front legs 'with claws' / front legs 'like paddles'; Q4. Has legs *Shinisaurus crocodilurus* Has no legs *Ophiophagus Hannah* b. scaly skin
	7	a. i. fresh mass is greater because it includes water ii. water content is very variable iii. data is more reliable since 'rogue' results can be discarded b. i. fresh mass increases from 0.46g to 0.64g ii. seed is absorbing water from the soil a. i. dry mass falls from 0.46g to 0.24g ii. food stores are being used up (e.g. in respiration) b. growth (cell division) / respiration / excretion / sensitivity
	Crossword	ACROSS: 1, reproduction; 6, protoctist; 9, sensitivity; 10, gene; 12, animal; 13, development; 16, respiration; 21, variation; 22, movement; 23, kingdom; 24, fungi; 25, energy. DOWN: 2, excretion; 3, Panthera; 4, HomoSapiens; 5, bacterium; 7, binomial; 8, Linnaeus; 11, organisation; 14, nutrition; 15, growth; 17, taxonomy; 18, plant; 19, species; 20, key.
7: CELLS, TISSUES and ORGANS	1	a. C (1, 2, 3), B (1, 2, 3), F (1, 4, 5, 6, 7), I (1, 4, 5, 8), G (1, 4, 5, 6, 7) b. i. red blood cell – transport of oxygen from lungs to tissues ii. lymphocyte – production of antibodies as part of immune response iii. macrophage – engulfing and destroying pathogens
	2	a. cell wall = A, cytoplasm = B, vacuole = E b. nucleus c. plant cell has chloroplasts / cellulose cell wall / large permanent vacuole – animal cell has none of these d. nucleus e. tissue f. no cytoplasm / no cell membrane
	3	a. A = nucleus, B = cell membrane b. photosynthesis (absorption of light energy to produce carbohydrates) c. plant cells have a cellulose cell wall / large permanent vacuole – animal cell does not d. $50 \times 20 = 1000$ i.e. one cell is $1000\mu m^2$. $1 mm^2 = 1000 \times 1000\mu m^2$. Therefore, 1000 plant cells could fit into $1 mm^2$.
	4	a. Cells listed from top to bottom: 1 – carries oxygen around the body; 2 – moves dust and bacteria up the bronchi; 3 – absorbs water and minerals from soil for the plant; 4 – transports water and minerals; 5 – contracts to cause movement within animals b. Heart is made up of more than one tissue e.g. muscle / nerve / epithelium

CHAPTER	QUESTION	ANSWER(S)
	Fill in the missing words	a. Cells, tissues, epidermis / xylem / phloem. Organ. Systems, excretory system. b. Specialized. Red blood cell. Division of labour. Nervous, endocrine. c. Palisade cell, chloroplasts. Leaf, epidermis, xylem.
8: MOVEMENT OF MOLECULES IN AND OUT OF CELLS	1	a. osmosis b. partially permeable (or differentially permeable) c. water has been absorbed d. it has the same water potential as the contents of cell e. add vacuole, and refer to Chapter 8, question 3
	2	a. concentration on x axis, percentage change on y axis / labels on axes / include negative change to mass of potato b. at crossover on x axis i.e. around 0.24 molar c. there is no change in mass as there is no net water gain or loss since cytoplasm and surrounding solution have the same water potential d. to allow calculation of a mean change (more reliable) e. to remove surface water which would affect the accuracy of the measurement of mass f. expected loss of 20% from 10 g suggests new mass of 8.0 g

	3	a. i. as raw material for photosynthesis / to dissolve soluble molecules for transport / to provide support through turgor
		   ii. salty soil solution has lower water potential than cell contents. Water leaves by osmosis down the water potential gradient, so cells die as less water available to them
		b. i. active transport
		   ii. growth rate might fall as active transport uses up energy, which might otherwise be used for growth
		   iii. nitrate – production of amino acids and proteins / magnesium – part of the chlorophyll molecule

	Fill in the missing words	Cytoplasm, partially permeable membrane. Swell, lower. Cell wall. Amino acids, diffusion, concentration gradient. Active transport, energy, against. Carbon dioxide, diffusion, photosynthesis.
9: BIOLOGICAL MOLECULES AND FOOD TESTS	1	a. A, F, G, H b. Dissolve sample in water / add equal volume of Benedict's Reagent / heat in a water bath / look for colour change from blue to orange-red

	2	a. 1: iodine solution – starch
		   2. Benedict's Reagent – simple sugar / glucose
		   3: alcohol – fat or oil
		   4: Biuret – protein
		b. Test 1 – straw-brown colour
		   Test 2 – blue
		   Test 4 – blue
		c. i. Milk / cheese / butter
		   ii. fish / meat / beans / cheese

	3	a. A b. B c. D d. E e. C

	4	a. mark for accuracy / labelling / key
		b. pie chart – easier to see the differences than when shown in a table
		c. i. water
		   ii. protein
		   iii. water
		   iv. fat
		   v. vitamins
		   vi. bones and teeth

	5	a. blue-black b. Biuret solution c. D d. as controls – to show that distilled water gives no colour change e. dissolve fat in alcohol / shake / pour suspension into distilled water / cloudy suspension indicates presence of fat f. cardiovascular disease (arteries become narrowed) / becoming obese – damage to joints

CHAPTER	QUESTION	ANSWER(S)
	Crossword	ACROSS: 2, water; 4, starch; 6, Biuret; 8, blood; 9, orange; 10, glucose; 11, cholesterol; 14, DNA; 15, condensation; 16, emulsion; 18, blueblack; 20, lipid; 21, vitamins; 22, haemoglobin. DOWN: 1, purple; 3, amino acid; 5, hydrolysis; 7, protein; 12, sucrose; 13, Benedicts; 17, milky; 19, urine
10: ENZYMES CONTROL BIOLOGICAL PROCESSES	1	PROTEINS; PATHWAYS; CATALYSTS; SPECIFIC; DENATURED
	2	a. temperature b. pH (use buffer solution), concentration of enzyme (use identically-sized pieces of potato), concentration of substrate (fix the concentration of hydrogen peroxide) c. rate of oxygen production – measure volume produced as the fluid is pushed around the manometer scale in a fixed period of time d. repeat the experiment with boiled potato pieces (enzymes would then be denatured) e. mean improves **reliability** of data (each individual value has less effect on the result). **Validity** is improved by carefully controlling variables so that there is only on input variable.
	3	a. temperature on x axis, oxygen consumption on y axis / curve represents the effect of temperature on enzyme action, with peak corresponding to the optimum temperature b. input = temperature, outcome = oxygen consumption; fixed might be: pH of environment, mass of maggots, age of maggots, food availability for maggots
	4	a. an enzyme is a protein which acts as a biological catalyst b. i. substrate – the stain / dirt to be removed – molecules fit onto enzyme's active site. These molecules are then broken down more quickly than they would be without the enzyme present. Breakdown would be able to go on at lower temperatures, thereby lowering the energy cost for the process. ii. enzymes are sensitive to temperature – too low and the molecules do not collide with the active site with enough energy; too high and the enzyme could be denatured so that the active sites would lose their shape iii. 35 – 40 °C: low energy cost but close to optimum temperature of enzyme c. microbes grown in fermenter / bioreactor under optimum conditions of temperature and pH. Microbes will produce more of the enzymes if the expected substrate (e.g. fats) is present. Collect microbes and crush to release enzymes; separate enzymes from the solution – dry.
	5	a. i. POTATO: 4.3, 6.3, 0.0, 0.0 LIVER: 8.0, 10.0, 0.0, 0.0 ii. sample / treatment on x axis, volume on y axis; is key suitable; bars should not be touching iii. potato has a lower output of oxygen – 4.3 compared with 8.0 cm^3 iv. sample B has potato in smaller pieces, so larger surface area of enzymes exposed to the substrate b. they are controls which demonstrate that an enzyme is responsible for the process, since boiling denatures the enzyme c. oxygen will relight a glowing splint
	6	a. it is fixed onto some form of support, such as a plastic grid b. oxygen concentration will fall as it reacts with glucose c. converts one form of 'energy' (e.g. peroxide concentration) into another form of energy (e.g. the display on the monitor)
	Crossword	ACROSS: 1, denaturation; 5, metabolism; 8, hydrogen; 9, cyanide; 10, substrate; 13, inhibitor; 15, protease; 16, catalyst; 17, and; 18, lock; 19, pepsin; 20, optimum DOWN: 2, temperature; 3, lipase; 4, amylase; 6, active; 7, key; 10, site; 11, activator; 12, catalase; 14, product.
11: FOOD AND DIET IN HUMANS	1	a. A b. A c. C d. D e. A
	2	a. 0.05 (i.e. 10 × 0.1 / 20) b. C c. it reduces it, by a factor of 4 (4 times as many drops are needed to decolourise the DCPIP) d. to prevent scurvy e. some leafy vegetables, potato

CHAPTER	QUESTION	ANSWER(S)
	3	a. i. Boy requires more as he is still growing: producing more cells, especially muscle, requires protein. ii. girl is menstruating so is losing blood each month: replacing the blood requires iron iii. the pregnant woman requires extra calcium for producing the bones of her developing child b. Vitamin C prevents scurvy, by helping the body to make fibres of an important protein needed for healthy gums and skin.
	4	a. Quorn provides good quantities of protein (for growth) and carbohydrate (as an energy supply) as well as fibre to help formation of the faeces, but low quantities of harmful fat and cholesterol. b. Quorn would provide 250 / 100 × 355 kJ of energy. Olive oil would provide 10 / 100 × 3600 kJ. Assume that the potato has a mass of 100g and so supplies 575 kJ. Total energy content is 1882.5 kJ.
	5	a. blue-black b. Biuret solution c. D d. as controls – to show that distilled water gives no colour change e. dissolve fat in alcohol / shake / pour suspension into distilled water / cloudy suspension indicates presence of fat f. cardiovascular disease (arteries become narrowed) / becoming obese – damage to joints
	Crossword	ACROSS: 1, scurvy; 3, salt; 4, calcium; 7, carbohydrate; 9, steroid; 10, protein; 11, fat; 13, respiration; 14, balanced; 15, iron; 16, membrane DOWN: 2, rickets; 3, sucrose; cholesterol; 6, baked beans; 8, malnutrition; 12, fibre
12: DIGESTION AND ABSORPTION	1	a. tongue b. A pushes food backwards, and mixes it with saliva from B. This makes the food more 'slippy' / better lubricated and so it is easier to swallow c. it closes of the trachea so that food does not enter the lungs d. show longitudinal and circular muscles – circular muscles contracting behind bolus and relaxing ahead of bolus e. it is long – up to 6m – and it is folded f. i. Glycogen ii. hepatic portal vein iii. hepatic vein iv. bile emulsifies fat (increases the surface area of fat globules), and has its effects in the duodenum / small intestine g. i. along the pancreatic duct ii. amylase (starch – glucose), lipase (fat – fatty acids and glycerol), protease (peptides – amino acids) h. E: reabsorbs water / synthesizes vitamins e.g. vit K F: stores faeces before defaecation
	2	a. A would be blue-black, B would be straw-brown b. in A the amylase had been boiled and so was denatured: the starch had not been digested. In B the starch would have been digested and so could not react with the iodine. c. test contents of A and B with Benedict's Reagent. Heating should make contents of B turn from blue to orange-red. d. i. Surface epithelium, capillary, lacteal ii. lining the small intestine (especially the ileum) iii. part X iv. to transport absorbed lipids / fatty acids
	3	a. i. molar, canine, incisor ii. X crushes food, Z bites off pieces of food before they are ingested b. calcium and vitamin D c. i. bacteria convert sugar in food to acid / acid eats away the enamel / bacteria (in plaque) can now enter the dentine ii. brush teeth carefully / reduce intake of sugary or acidic foods / drink water treated with fluoride
	4	a. lipase b. fatty acids c. i. Falls by 1 pH unit over 4 minutes, so an average of 0.25 ph units per minute ii. bile increases the surface area of the fat globules so that the lipase enzyme can work more quickly

CHAPTER	QUESTION	ANSWER(S)
	5	a. i. protease ii. amino acids b. pH on x axis, rate on y axis / labels include quantity and units / title could be 'The effect of pH on the activity of a protease' / curve should have an optimum at pH 2.0 c. stomach d. too large to be absorbed / cross the lining of the ileum
	Matching terms	A-12; B-22; C-21; D-17; E-3; F-13; G-19, H-1; I-20; J-18; K-2; L-14; M-9; N-16; O-6; P-8; Q-7; R-5; S-4; T-15; U-11; V-10

13: PHOTOSYNTHESIS AND PLANT NUTRITION	1	a. either by adding / removing more lamps or by moving the lamp to different distances from the plant
		b. remove the lamp / remove plant from the beaker
		c. temperature / carbon dioxide concentration in the water / species of plant / wavelength of light
		d. i. Light intensity on x axis / rate on y axis / labels to include quantities and units / points joined with smooth curve / maximum reached between 60 – 70 units
		    ii. 26 / 27 bubbles (cannot have half bubbles!)
		    iii. rate of photosynthesis is proportional to light intensity up to about 60 units of light intensity. Beyond this, light intensity has less effect as some other limiting factor is 'controlling' the rate of photosynthesis (e.g. carbon dioxide concentration)
		    iv. less carbon dioxide / more oxygen
		    v. measure volume of gas given off, as bubbles may not all be of the same size.
		e. input variable would be colour / wavelength (vary by using different filters or coloured papers), outcome variable would be rate of photosynthesis, fixed variables would be temperature, carbon dioxide concentration and MUST include light intensity

	2	a. carbon dioxide + water → (have chlorophyll and light above arrow) → glucose + oxygen b. photosynthesis c. rises when light intensity increases, as more photosynthesis possible / falls as light intensity falls and sugar is used up in respiration faster than it can be produced d. light intensity was lower that day (e.g. it was cloudy) e. raises temperature (affects enzymes of photosynthesis) / provides more carbon dioxide as fuel is burned

	3	a. i. A – upper epidermis, B – palisade mesophyll, C – spongy mesophyll, D – guard cell of stomata, E – xylem
		    ii. palisade mesophyll – B
		    iii. xylem – E
		b. as sucrose (in the phloem)
		c. i. starch
		    ii. add iodine solution – blue-black colour is a positive result
		    iii. could be respired to release energy for growth / could be converted to cellulose in cell walls
		    i. 1.4 mm (measured thickness is 70 mm, scale has 5 mm representing 0.1 mm, 70/5 = 14 and 14 × 0.1 = 1.4 mm
		    ii. approx 0.4 mm – same reasoning.

	4	a. light intensity / temperature / volume of water available / species of plant / age of plant at start
		b. i. stem and roots are smaller / some leaves yellowed / plant cannot manufacture proteins and other compounds needed for growth
		    ii. leaves pale green or yellow / plant cannot produce chlorophyll as magnesium is a part of this molecule
		c. i. plants had absorbed / removed the nitrate ions; nitrates had been washed out of the soil; nitrates converted to nitrogen gas by denitrifying bacteria
		    ii. plant nitrogen-fixing plants (e.g. peas and beans) / add nitrate fertilizer / add organic compost which bacteria could break down to nitrate
		d. some may capture insects and obtain nitrates from insects' bodies / some may have root nodules containing nitrogen-fixing bacteria

	Matching terms	A-4; B-10; C-15; D-1; E-13; F-8; G-5; H-14; I-12; J-6; K-2; L-9; M-11; N-7; O-3

CHAPTER	QUESTION	ANSWER(S)
14: PLANT TRANSPORT	1	a. key point here is that RATE is proportional to 1 / time taken. Rates are 2.0, 0.63, 10, 0.59, 0.40. On bar chart, the bars should not touch / environmental condition on x axis, rate on y axis. b. At high light intensity plants will photosynthesize so stomata are open – water vapour can leave easily so rate of bubble movement is quite high. c. (2) air is humid so water potential gradient is not steep; (5) low humidity means that some water loss can take place from the few stomata that are open in the dark (little photosynthesis so no light available) d. (3) wind removes moist envelope of air from leaf surface so that water potential gradient is steep; (4) in the dark most of the stomata are not open as no photosynthesis can take place, so even windy conditions will not lead to a high rate of bubble movement.

	2	a. i. lower surface – look at leaf C compared with B
		   ii. yes – look at D compared with C
		b. i. X is stomatal pore, Y is guard cell
		   ii. lower surface of leaf has more stomata, so more water vapour can be lost. Vaseline blocks these stomata so loss of mass from B is very much reduced.

	3	a. stomata
		b. more complex diagram of simplified one on p. 60
		c. i. evaporation
		   ii. diffusion
		   iii. helps in cooling plant in direct sunlight, BUT means that large volumes of water must be replaced by the roots
		d. leaf X (18, 18,000), leaf Y (8, 8,000) – there are 1000 mm² in an area 5 cm × 2 cm
		e. leaf Y – fewer stomata from which water vapour can be lost so less need to replace this lost water
		f. thick, waxy cuticle / hairs on leaves / leaves may be rolled with stomata on inside surface

	4	a. i. 1: stomata are kept in a humid environment so no water potential gradient down which water vapour will move; 2: water cannot move through the waxy layer
		   ii. 1: can absorb water from great depths in the soil; 2: fleshy stem has cells which can store large volumes of water when water is available
		b. less opportunity for light absorption for photosynthesis, so fewer materials available for growth
		c. i. osmosis / photosynthesis / evaporation (transpiration / diffusion)
		   ii. water potential of soil solution / light intensity / temperature, wind speed

	Fill in the missing words	Osmosis, hairs. Surface area, ions, nitrate. Diffusion, active transport. Support, solvent, photosynthesis. Xylem. Phloem, vascular.
15: TRANSPORT IN ANIMALS	1	a. Jack – more red blood cells to help when oxygen concentration is very low b. James – more white blood cells for defence c. Jack – has more platelets (which increase blood clotting) and red blood cells (which would increase the viscosity of the blood) d. Julian – has fewer red blood cells which are responsible for the transport of oxygen. Less oxygen means cells would be less able to release energy, and sufferer would be fatigued. e. it has removed age and gender as possible variables from the investigation
	2	a. C b. C c. A d. A e. B f. B g. B h. B i. B j. C
	3	a. 120 000 (male and female) b. 245 000 c. 120 000 / 500 000 = 24% d. pie chart makes it more obvious which is the largest contributor to deaths

CHAPTER	QUESTION	ANSWER(S)
	4	a. i. Hepatic vein – vena cava – right atrium – right ventricle – pulmonary artery – pulmonary vein – left atrium – left ventricle – (aorta) – coronary artery ii. pulmonary vein – left atrium – left ventricle – aorta – hepatic artery – hepatic vein – vena cava – right atrium – right ventricle – pulmonary artery b. i. Plenty of exercise (preferably aerobic) / diet low in saturated fats ii. diet high in salt / smoking (accept mixture from i. and ii.)
	5	a. i. 0.8 seconds ii. 60 seconds / 0.8 seconds per beat = 72 beats per minute b. i. maximum in left ventricle = 130 mm mercury, so maximum in right ventricle = 130 / 5 = 26 mm mercury ii. left wall is much more muscular c. pressure in ventricle is greater than in aorta so blood can flow out of heart into the aorta d. X: left atrioventricular (bicuspid) valve – prevent backflow from ventricle to atrium as pressure in ventricle rises; Y: left semilunar valve – prevent backflow from aorta to left ventricle as pressure in ventricle falls
	6	a. i. oxygen / glucose / amino acids ii. carbon dioxide b. i. muscle ii. it will contract (the fibres will shorten) iii. they are put under pressure so will be forced out along the aorta c. i. diet high in saturated fats and / or cholesterol / low level of exercise / smoking ii. shade areas after the blockage (i.e. downwards on the diagram) d. they have thin walls / wide lumen so that they can carry large volumes of blood / they have valves to help return of blood at low pressure to the heart
	Crossword	ACROSS: 1, cholesterol; 3, pacemaker; 4, muscle; 7, blood pressure; 9, arteries; 10, coronary; 11, semilunar; 15, tendons; 16, ventricle; 18, left; 19, pulmonary; 21, atrium; 22, veins DOWN: 1, capillaries; 2, valve; 5, smoking; 6, vena cava; 8, exercise; 11, diastolic; 13, systolic; 14, aorta; 17, bypass; 19, heart rate; 20, right
16: DEFENCE AGAINST DISEASE	1	a. phagocyte / macrophage b. phagocytosis c. it can detect an antigen (foreign protein) on the surface of the bacterium d. digestive enzymes are added e. cholera, pneumonia, syphilis, gonorrhoea, TB and many others f. antibiotics can be prescribed
	2	a. a product which can stimulate the immune response b. a protein made by a B-lymphocyte which can recognise and bind to an antigen c. it is larger, quicker and lasts for longer d. the vaccine is less likely to cause unwanted side-effects if given in two smaller doses. The second vaccination stimulates the production of more memory cells. e. A: active; B: active; C: passive; D: passive; E: passive
	3	a. i. A – 3; B – 1; C – 4; D – 2 ii. AIDS – unprotected sex / sharing needles for drug users; cholera – drinking infected water; athlete's foot – contact with fungal spores e.g. on a towel; malaria – being 'bitten' by an infected *Anopheles* mosquito b. i. these are the nutrients for the growth of the bacterial culture ii. to sterilize it (make sure that there are no microbes present in it) iii. bactericidal would show bacteria being killed (so population falls to zero), bacteriostatic would show population stable, no longer increasing iv. any bacterial diseases e.g. *Salmonella*; infection; cholera; TB; syphilis
	4	a. Penicillin (or any acceptable alternative) b. 1990–1993 c. Some bacteria mutate/become resistant. Non-resistant bacteria killed/prevented from reproducing. Resistant bacteria multiply so whole population becomes resistant. Antibiotic now less effective. d. i. less effective / too expensive in higher doses ii. different antibiotic / better sterilisation techniques
	Fill in the missing words	Blood loss, pathogens, disease. Platelets. Proteins. Fibrinogen, fibrin. Red blood cells. Anaemia.

CHAPTER	QUESTION	ANSWER(S)
17: BREATHING AND GASEOUS EXCHANGE	1	A – d; B – c; C – a; D – e; E – b
	2	ENERGY; RESPIRATION; OXYGEN; CARBON DIOXIDE; SURFACE AREA; THIN; MOIST; VENTILATION; BLOOD; AMOEBA; SPIRACLES; TRACHEOLES; GILLS
	3	a. intercostal muscles contract to move the ribs upward and outwards. The diaphragm contracts and moves downwards. These two movements increase the volume of the chest (and the lungs) so the pressure inside the lungs falls: this causes air to move in from the atmosphere. b. i. The total volume of air breathed increases steadily up to 5.5% carbon dioxide, then it increases very rapidly to a maximum of 55 litres ii. because the total volume depends on the volume per breath (tidal volume) as well as the breathing rate
	4	a. cigarettes per day / period since smoking on x axis, annual death rate on y axis; check the scales on the axes to make full use of graph grid; check key to distinguish two curves b. 7.5 c. 4 years d. nicotine e. carbon monoxide in tobacco smoke reduces the ability of haemoglobin to transport oxygen, so lips look less red as less oxyhaemoglobin is formed f emphysema / bronchitis / lung cancer / bladder cancer / mouth cancer
	5	a. before: 10 after: 13 b. 0.9 dm³ per minute c. The student is still repaying an oxygen debt. During the exercise the muscles produce lactic acid – getting rid of this requires oxygen even when no further exercise is going on.

	6	a. i. the random movement of molecules down a concentration gradient (until an equilibrium is reached)
		   ii. there is a higher concentration of oxygen in the alveoli – the gas moves down a concentration gradient
		   iii. they are thin, moist, have a large surface area, are well-ventilated and are close to a good blood supply
		b. i. the concentration gradient from alveoli to blood would be less steep, so it would be more difficult for oxygen to be taken up
		   ii. more red blood cells increases the oxygen-carrying ability of their blood. This provides more oxygen for respiration, and so more energy can be released and the athlete can perform better.

	Fill in the missing words	Pulmonary. Respiration, energy. Carbon dioxide, hydrogencarbonate. Capillaries, thin, surface area. Alveoli. Carbon dioxide, oxygen. Diffusion. Pulmonary, left atrium.
18: RESPIRATION	1	Sensitivity / locomotion / growth / reproduction / nutrition / growth and development
	2	a. activity on x axis, value on y axis / keep bars separate / label bars correctly b. 5150 kJ (450 + 1000 + (4x650) + 1100) c. to keep heart beating / digestion operating / maintain body temperature / keep breathing d. water is denser than air, so resistance to movement is greater

	3	a. i. 50°C (79–29)
		   ii. 8.4 kJ per g
		   iii. should be in the form of a bar chart, bars separate and clearly identified / type of food on x axis, energy value on y axis
		   iv. fat (closest to value for energy content of seed)
		b. Benedict's Test: crush seed and suspend in water / heat sample with equal volume of Benedict's Reagent / look for colour change: blue to orange-red if reducing sugar is present

CHAPTER	QUESTION	ANSWER(S)
	4	a. i. carbon dioxide ii. respiration iii. cell division / movement / synthesis of molecules / maintaining body temperature b. absorbed onto crystals of soda lime (potassium or sodium hydroxide), or could be passed through limewater c. size reflects the number of respiring cells. Same size is removing 'number of cells' as a variable in this experiment d. mouse feeds on radioactive glucose / respires glucose and so releases radioactive carbon dioxide / carbon dioxide reacts with lime water to form calcium carbonate / calcium carbonate solid is collected on filter paper
	5	a. F; b. T; c. F; d. T; e. T; f. T; g. F; g. T; h. T; i. T; j. F
	Fill in the missing words	Living cells. Glucose, oxygen. Energy, aerobic. Growth, movement, heat.
19: EXCRETION AND OSMOREGULATION	1	a. total water gain = 2700 cm³ / water loss = 450 + 500 + 150 + loss in urine. Therefore loss in urine = 2700 − 1100 = 1600 cm³ b. Receptors in brain measure water potential of blood. Changes in water potential alter the amount of ADH (AntiDiureticHormone) released. This hormone then controls the amount of water 'saved' by the kidney and so not lost in the urine. This demonstrates negative feedback, since a reduction in water intake eventually causes greater saving of water and so 'cancels out' the original change. c. i. it does not depend on the availability of a kidney for transplant / it should not challenge the patient's immune system / drug treatment can go on through the dialysis fluid entering the body ii. it is inconvenient – the patient must spend long periods on the machine / long-term treatment is very expensive
	2	i. The protein molecules are too large to cross the filtration membrane and enter the urine. ii. Glucose molecules can cross the filtration membrane, but it is possible for them to be selectively reabsorbed and returned to the blood.
	3	a. i. Excretion is the removal from the body of toxic materials, the waste products of metabolism. ii. Egestion is the removal from the body of food materials that have not been digested and absorbed from the gut. b. i. in the liver ii. from excess amino acids (by the process of deamination) c. Q = renal artery; R = vena cava; S = ureter; T = urethra d. glucose (X); red blood cells (X); salts (√); water (√)
	4	a. Urea is still being formed but it is not being excreted. b. It will take more time before the urea concentration reaches dangerous levels. c. The patient may be treated with drugs which suppress the activity of their immune system, so that the body's defences no longer attack the donor kidney / the donor kidney may be treated so that it has no antigens on its surface and so it is not recognised by the host defences.
	Crossword	ACROSS: 2, urine; 3, urea; Henle; 5, bladder; 8, filtration; 9, renal artery; 10, sphincter; 13, excretion; 14, plasma; 15, glucose; 16, Bowmans, 18, urea; 19, kidney; 20, dialysis DOWN: 1, nephron; 2, urethra; 3, ureter; 6, renal vein; 7,diuretic; 11, hormone; 12, amino acids; 14, protein; 17, anti
20: THE NERVOUS SYSTEM AND COORDINATION	1	a. A (accept E); b. C; c. E; d. E; e. D; f. B
	2	a. 4 b. 4 c. 80 mg corresponds to 5 units of alcohol. 3 cans of cider plus 1 glass of wine = 10.6 units of alcohol. This corresponds to 175 mg per 100 cm³, so the person would be 95 mg per 100 cm³ over the legal limit d. reduces coordination / increases reaction time / affects judgement of distance e. Five and a half to six hours f. cirrhosis

CHAPTER	QUESTION	ANSWER(S)
	3	a. i. Road traffic accidents ii. 5/20 – 25% iii. meningitis is an infection that can affect the spinal cord b. she may have damaged her sensory nerves, or the dorsal root of the spinal nerve (so she can't feel things) but the motor nerves are unaffected (so she can write) / senses from the feet may enter the spinal cord at a different level to motor impulses leaving the spinal cord for the hands
	4	a. A = sensory neurone; B = grey matter; C = relay neurone; D = motor neurone; E = synapse b. Show gap / neurotransmitters c. Caffeine: stimulant – allows neurotransmitter to act for longer; marijuana: sedative – neurotransmitters cannot cross the synapse and continue an impulse; Heroin: sedative – copies the effect of a natural sedative chemical; cocaine – stimulant: allows neurotransmitter to act for longer; nicotine: stimulant – copies the effect of a natural stimulant chemical
	5	a. i. 73, 75, 79, 79, 83, 87 ii. number of cups on x axis, rate on y axis – probably best shown as a bar chart iii. yes – coffee appears to increase the heart rate iv. cups contained the same volume of coffee / coffee always the same 'strength' / volunteers had similar background (e.g. age / gender / basic heart rate) b. examine heroin use by all women giving birth / look for relationship between birth weight and heroin use by mother. This is necessary since it is unethical to give heroin to mother and see what happens to the birth weight of her child!
	6	a. relay neurone – transfers nerve impulse from sensory to motor neurone b. synapse c. i. as an electrical impulse (wave of depolarisation) ii. as molecules of neurotransmitter d. spinal (knee jerk / withdrawal); cranial (pupil reflex / accommodation reflex); conditioned (salivation when thinking of tasty food) e. i. it would be lost ii. it would be unaffected f. no – the dorsal root is not itself sensitive (the receptor is the part of the reflex arc that 'detects' the stimulus) g. C = axon; D = myelin sheath; E = cell body; F = (accept nodes of Ranvier); G = dendrite. Two differences from motor neurone: cell body nearer middle of neurone / neurone attached to receptor cell / impulses travel towards the CNS
	Matching terms	A-5; B-9; C-1; D-10; E-2; F-3; G-6; H-4; I-7; J-8
21: THE EYE AS A SENSE ORGAN	1	a. i. Before: small pupil, large iris; after: large pupil, small iris ii. pupil in middle / iris around pupil / sclera = 'white' of eye b. light is detected on retina, motor impulse to radial muscle of iris, muscle contracts, iris becomes narrower, pupil larger c. shapes rely on black and white vision, which is possible because rod cells operate well at low light intensity. Colour depends on cones, which do not operate well at low light intensity.
	2	a. cornea and lens b. A would contract, ligaments would loosen, lens would become shorter and fatter, light would focus more onto retina
	3	A-5; B-9; C-1; D-10; E-2; F-8; G-4; H-3; I-6; J-7
	4	a. A: ciliary muscle; B: contracts to make pupil smaller in bright light b. i. voluntary: under conscious control; antagonistic: with opposite effects ii. the eye would look to the right iii. D would contract and C would relax c. cornea – aqueous humour – pupil – lens – vitreous humour d. rods: monochrome / low intensity – throughout retina, especially at the edges. Cones: colour / high intensity – at the centre of the retina (yellow spot)

CHAPTER	QUESTION	ANSWER(S)

	5	a. i. light receptors convert light energy to electrical energy, as nerve impulses
		     ii. rods – sensitive to light at low intensity / black-and-white vision;
		         cones – sensitive to light of high intensity / colour vision
		b. a neurone / relay neurone – has many dendrites to connect to other cells
		c. cornea – aqueous humour – pupil – lens – vitreous humour
		d. i. biceps contracts / triceps relaxes / forearm is pulled so that the elbow flexes
		     ii. optic nerve (sensory nerve) – brain – spinal cord – motor neurone – muscles in arm

	Crossword	ACROSS: 2, retina; 5, balance; 6, eye; 7, coughing; 9, sight; 11, cones; 12, ligaments; 15, reflex; 18, rods; 19, nose; 22, touch; 25, thermo; 26, stretch; 28, ciliary muscle; 31, stimulus; 32, skin; 33, iris; 34, choroid DOWN: 1, receptor; 3, tongue; 4, lens; 8, fovea; 10, hearing; 13, integration; 14, taste; 16, smell; 17, auditory; 20, optic; 21, ear; 23, cornea; 24, blind spot; 26, sensory; 27, tears; 29, pain; 30, pupil
22: HORMONES AND THE ENDOCRINE SYSTEM	1	Insulin / oestrogen / progesterone
	2	a. The cattle will gain weight more quickly, produce more milk and will have less of their 'meat' as fat. b. an increase in the number of infertile cattle c. Humans may receive BST from the meat, and this may affect the human's growth. d. Otherwise the BST could be digested in the intestine before it has been absorbed. e. Chickens would appear to be a cheap food source for the body builders, but the chickens would retain oestrogen in their tissues. The body builders might receive this oestrogen, and since this is a female hormone the male body builders become feminized.
	3	a. A – 5; B – 1; C – 8; D – 6; E – 2; F – 7; G – 3; H – 4 b. Adrenaline can therefore prepare working muscles for longer periods of greater activity, and can make sure that the brain makes the decisions necessary to deal with the issue (e.g. 'shall I fight or shall I run away?')
	4	a. Time on *x* axis, glucose and insulin levels on *y* axis / check appropriate scales for the two dependent variables / check that 'activity' periods are indicated b. Glucose level falls, as glucose is removed to muscles for respiration to release energy. The insulin level falls as it is no longer necessary to 'instruct' the liver to remove glucose and store it as glycogen. Half an hour later the insulin level rises following the increase in glucose level, and the need to remove glucose for storage. c. Glucagon (increases breakdown of glycogen to glucose) / adrenaline (increases availability of glucose for respiration) d. i. This reduces the likelihood that any unreliable single result will affect the overall result ii. it can be assumed that they will have a normal glucose / insulin response e. 1 hour
	5	a. a chemical, produced by an endocrine organ, released into the blood where it travels to affect the activity of a target organ b. a rise in blood glucose level is detected by the Islets of Langerhans in the pancreas. These glands release insulin. Insulin increases the uptake of glucose by the liver, and the storage of excess glucose as glycogen. As a result the blood glucose levels return to normal. This is an example of negative feedback. c. insulin would be digested to peptides and amino acids, and so its function would be lost. (Insulin can be wrapped in a gel coat to partially overcome this problem)

	6	a. i. Insulin increases the uptake of glucose by the liver, and the storage of excess glucose as glycogen. As a result the blood glucose levels return to normal.
		     ii. Insulin increases the uptake of glucose by the liver, and the storage of excess glucose as glycogen. As a result of too much insulin the blood glucose levels fall below normal.
		     iii. glucose (sugar) is removed from the blood by the working tissues, so that it can be respired and release energy
		b. sugar is absorbed more quickly, as starch has to be digested before it can provide sugar.
		c. a fall in blood glucose level is detected by the Islets of Langerhans in the pancreas. These glands now release less insulin. As there is less insulin there is a fall in the uptake of glucose by the liver, and the storage of excess glucose as glycogen. As a result the blood glucose levels return to normal. This is an example of negative feedback.
		d. in negative feedback, a change from normal conditions (e.g. an increase in blood sugar level) sets off a process to cancel out the change (in this case, the release of insulin from the pancreas)

CHAPTER	QUESTION	ANSWER(S)
	Fill in the gaps	Hormones, endocrine organ, blood, target organ. Adrenaline, trachea.
23: SENSITIVITY AND MOVEMENT IN PLANTS	1	a. growth, stimulus b. gravity c. positive, photosynthesis
	2	C; it is the tip which produces the growth hormone, auxin, responsible for the phototropism
	3	a. concentration on x axis, curvature on y axis / correct labels with quantity and unit / check that scale is correct / accuracy of plot b. 12 degrees c. positive phototropism d. auxin
	4	a. Shoot B has not increased in length. Auxin cannot reach shoot B as the tip has been removed. Auxin is needed for growth, so no auxin means no growth. b. i. Auxin has moved down the shoot and towards the non-illuminated side of the shoot ii. R c. Plant hormones can make flowers act as though they have been fertilized, and so produce seedless fruits / plant hormones can be used as selective weedkillers / plant hormones can be used to control growth of hedges / plant hormones can coordinate the formation of fruits so that harvesting is more efficient

	5	a. a suitable temperature for enzyme action / availability of water / availability of oxygen for aerobic respiration
		b. so that light could reach the shoot tip
		c. i. shoot would be growing upwards after a short 'horizontal' length / root would be growing downwards after a short 'horizontal' length
		   ii. root: positive geotropism – auxin pulled to bottom surface of horizontal root – cells grow less – root curves downwards. Shoot: positive phototropism – auxin diffuses to bottom of horizontal shoot – cells grow more – shoot curves upwards. Note that auxin has a different effect on the root cells than the shoot cells
		   iii. the continuous rotation means that the seedling was experiencing light and gravity stimuli from continuously differing directions

	Matching terms	A-5; B-7; C-6; D-9; E-3; F-10; G-4; H-1; I-8; J-2

25: REPRODUCTION IN PLANTS	1	a. i. pollination
		   ii. bee picks up pollen from anther of flower A, then transfers it to stigma when visiting flower B
		b. i. fertilization
		   ii. the ovary
		c. they will be similar because they are receiving genes from the same two parents, but not identical because the production of gametes involves meiosis (random assortment of chromosomes) and fertilization involves random fusion of gametes.

	2	a. M : sum of lengths / 10 = 247mm; N = 132mm b. 247 – 132 = 115mm c. the plants may be genetically different / there may be different nutrients or water levels in the soil d. This method of reproduction copies the genetic make-up of a plant and so the resistance to the weedkiller could be transmitted to offspring.
	3	a. sexual involves two parents and there can be very great variation between offspring / asexual has only a single parent and there is no variation between the offspring b. sexual – variation means that new environmental challenges can be met, BUT two parents are needed and many of the offspring may not be adapted to their environment / asexual – no variation means that offspring are suited to an unchanging environment BUT may not be adapted if environment changes in any way, only one parent needed means that this is a good method of colonising new environments where very few individuals may be found c. bud, which is close to ground, undergoes mitosis (copying division) and so grows out sideways; where stem touches the ground it forms roots (adventitious roots), then runner can break as a new plant has now been formed

CHAPTER	QUESTION	ANSWER(S)

	4	a. 2: stamens are almost hidden – stamens are obvious
		    3: five petals – six petals
		    4: style is longer – style is shorter
		    5: anthers large and elongated – anthers small and spherical
		    6: sepals are visible – sepals not visible
		b. i. P produces male gametes (pollen), Q receives pollen grains after pollination
		    ii. reproduction in flowers involves fertilization, which is part of sexual reproduction
		c. petals are large / and colourful / stamens do not hang out of flower / stigma does not hang out of flower / stigma is not branched to increase surface area

	Crossword	ACROSS: 4, stigma; 7, fruit; 9, hermaphrodite; 10, wind; 12, seed; 13, anther; 15, nectar; 17, ovary; 18, dispersal; 19, pollination; 20, insect; 22, variation; 23, filament; 24, tuber DOWN: 1, egg; 2, competition; 3, sepal; 5, asexual; 6, fertilization; 8, runner; 11, petal; 14, germination; 16, pollen; 21, style
26:GERMINATION AND PLANT GROWTH	1	a. A = testa (seed coat); B = cotyledon; C = radicle; D = plumule; E = embryo b. FAT : reagent – alcohol – add alcohol to crushed cotyledon / shake / pour suspension into distilled water / cloudy suspension indicates that fat is present STARCH : reagent – iodine solution – add iodine solution to crushed cotyledon – blue-black colour indicates that starch is present

	2	a. B and D – A has no water, C has no oxygen. B has both oxygen and water, at the correct temperature – the black cover is irrelevant since light is not a requirement for germination. D also has oxygen, water at an appropriate temperature
		b. i. temperature on x axis, percentage germination on y axis / labels on axes include quantities and units / plot is smooth curve with maximum at 35°C
		    ii. shape indicates that enzymes are involved – the shape indicates an optimum temperature typical of enzymes
		    iii. the hole means that it is easier for oxygen and water to enter the seed, and so germination will be quicker

	3	a. i. starch
		    ii. protein
		b. enzymes
		c. these are the areas of cell division, and amino acids are needed to produce the proteins needed for the building of new cells
		d. oxygen, water, an appropriate temperature for the action of enzymes
		e. i. positive phototropism
		    ii. this means that the shoot, with leaves, will be in the best position to absorb light for photosynthesis

	4	a. A – NO; B – YES; C – NO; D – YES
		b. i. check accuracy of plot
		    ii. 20
		    iii. 30

	5	a. oxygen and an appropriate temperature for the activity of enzymes
		b. i. In dish A, most of the seeds have germinated, with long radicles already showing the growth of root hairs. In dish B, just over half of the seeds have begun to germinate, but the radicles are short and none of them has root hairs.
		    ii. labels should include: testa, radical and root hairs
		c. i. some factor in tomato juice inhibits the germination of tomato seeds
		    ii. use the juice from the tomato seeds, and repeat the experiment but with seeds from other species.

	Crossword	ACROSS: 3, fertilization; 4, amylase; 5, embryo; 7, testa; 8, radicle; 9, ovule; 10, cotyledon; 13, flower; 15, plumule; 16, enzymes; 17, water DOWN: germination; 2, gamete; 6, oxygen; 11, dormancy; 12, fruit; 14, lipase; 15, pollen

CHAPTER	QUESTION	ANSWER(S)
27: HUMAN REPRODUCTION AND GROWTH	1	a. A: sperm duct / vas deferens; B: urethra; C: testis; D: scrotum / scrotal sac; E: penis b. testis / C c. prostate gland / F d. testis / C e. urethra / B f. sperm duct / A
	2	ovulation – ejaculation – fertilization – implantation – development – birth
	3	a. A = oviduct / fallopian tube; B = ovary; C = womb; D = uterus / wall of uterus; E = cervix; F = vagina / birth canal b. X in oviduct, Y in lining of uterus c. Diaphragm covers cervix, pill prevents release of eggs from ovary / ovulation

	4	a. i. carbon dioxide, urea
		  ii. villi give a large surface area, very close to mother's blood supply
		  iii. gut / intestine: will absorb digested food; lungs: will carry out exchange of carbon dioxide and oxygen
		b. i. provides a cushion for the developing fetus
		  ii. Downs syndrome (Trisomy-21)
		c. i. age of mother on x axis, frequency on y axis / bars can be touching / check accuracy of plot
		  ii. percentage difference = change / original x 100 i.e. percentage difference = $(27.2 - 4.9)/4.9 \times 100 = 455\%$
		  iii. advantage: prevent 'hot flushes' / limit deterioration of bones; disadvantage: tenderness of breasts / irregular menstrual bleeding / nausea

	5	a. i. a chemical, released from an endocrine organ, carried in the blood to have an effect on a target organ ii. in the bloodstream b. the green light gives an indication that there has not been a recent ovulation, so there is no egg cell to become fertilized c. oestrogen controls the repair of the lining of the uterus; luteinising hormone stimulates ovulation and controls the development of the corpus luteum after ovulation d. the timing of starting a family can be controlled until parents are ready (e.g. until mother is healthy following an illness) / unwanted pregnancies can be avoided / some people may use this contraception to become sexually promiscuous / there are side-effects of hormonal contraception, including increased possibility of dangerous blood clotting
	6	a. head: relative size of forehead decreases / ears become visible / mouth becomes visible; foot: separate toes become visible b. i. A: 8 mm; B: 94 mm ii. 12X (actually 11.75X)
	Matching pairs	A-13; B-9; C-8; D-1; E-3; F-18; G-15; H-2; I-10; J-20; K-5; L-6; M-7; N-4; O-17; P-11; Q-14; R-12; S-16; T-19
28: DNA AND CHARACTERISTICS	1	a. nucleus b. white c. B d. identical twins e. check the DNA fingerprints of individuals which are to be bred together. Look for the greatest difference in DNA fingerprints – this will ensure the greatest genetic variation in the offspring.

CHAPTER	QUESTION	ANSWER(S)
	2	a. Letters in top line : T,C, T; SHORT STRAND – (G); letters in bottom line: C, T, A, C, A; SHORT STRAND: C, T b. A – adenine, G – guanine, C – cytosine, T – thymine c. Mitosis is copying division, and replication provides two identical copies of the cell's DNA. d. gene e. i. Methionine – tyrosine – glycine – alanine – histidine ii. TAC – ATG – CCG – CCG – TA iii. methionine – tyrosine – proline – proline iv. mutation sickle cell anaemia / Huntington's disease / cystic fibrosis vi. the change may enable the organism to adapt to a new environment, e.g. to resist a particular disease
	3	a. A = DNA, B = nuclear membrane, C = messenger RNA, D = cytoplasm, E = amino acid, F = transfer RNA, G = ribosome b. i. Amylase / lysozyme ii. keratin iii. haemoglobin iv. pepsin
	4	a. meiosis b. gamete has half of the total DNA (the haploid number of chromosomes) c. DNA (deoxyribonucleic acid) d. top half : G, C, left strand: G, A; right strand A – T, G – C, A – T) e. mutation f. X-radiation, UV radiation g. sickle cell anaemia / Huntington's disease / cystic fibrosis
	5	a. i. Mitosis: amount of DNA doubles and then returns to normal amount as cell divides into two ii. DNA codes for the production of proteins, which give the cell its characteristics. b. mutation c. Isolate the gene to be cloned and insert it in a plasmid using a restriction enzyme and a ligase. Then insert the plasmid into a bacterium and grow it: this bacterium will carry the gene that was inserted.
	Matching terms	A-3; B-7; C-1; D-13; E-10; F-14; G-2; H-4; I-12; J-5; K-8; L-6; M-9; N-11
	Crossword	ACROSS: 2, amylase; 4, protein; 5, genotype; 7, mutagen; 10, chromosome; 13, messenger; 14, bases; 15, amino acid; 16, replication; 18, phenotype; 19, haemoglobin; 20, gene DOWN: 1, bacterium; 3, mutation; 6, nucleus; 8, translation; 9, ribosomes; 11, radiation; 12, double helix; 17, allele
29: CELL DIVISION AND THE HUMAN LIFE CYCLE	1	a. a string of genes, found in the nucleus of a cell b. i. 46 ii. 23 iii. 0 iv. 0 c. meiosis d. mitosis
	2	a. 46, 23, 23, 46, 46 b. it suggests that egg and sperm are from the same adult c. in the reproductive organs ovary in female and testis in male d. They must become haploid, because they fuse to form a zygote. If gametes were not haploid, zygotes would be tetraploid (4n) and the number would double with every generation.
	3	a. to make it easier to squash the tip and separate the cells b. to separate the cells c. to stain the chromosomes d. A, B, C, D, E, F e. show centromere, two chromatids f. 8 g. 4
	4	a. genes, and some factor in the environment b. when they contact another cell nearby c. mitosis d. bone marrow / skin / lining of gut / liver e. a mutagen f. surgery, chemotherapy, radiotherapy

CHAPTER	QUESTION	ANSWER(S)
	Matching pairs	A-7; B-5; C-8; D-6; E-1; F-9; G-2; H-10; I-3; J-4
30: PATTERNS OF INHERITANCE	1	a. phenotype b. allele c. Mendel d. gene e. genotype f. heterozygote g. dominant
	2	a. Rr, Rr; gametes would be R, r and R, r; A = RR (normal), B = Rr (carrier), C = Rr (carrier), D = rr (with cystic fibrosis) b. ¼ (1 in 4, 0.25) c. white blood cell – red blood cell does not have a nucleus and so cannot provide DNA for the probe to detect
	3	a. A – male, haemophiliac; B – female, carrier b. $X^N X^n \times X^N Y$ Gametes X^N X^n X^N Y Offspring $X^N X^N$, $X^N X^n$, $X^N Y$, $X^n Y$ – a haemophiliac son c. i. The female must receive the 'n' allele from *both* parents. ii. Girls could bleed to death when they menstruate.
	4	a. Gametes will be N, S and N, S A = NN, D = NS, C = NS, D = SS b. Low ability to transport oxygen, so less aerobic respiration so less energy c. malaria d. mutation e. X-radiation, U.V. radiation
	5	a. i. recessive – each parent must have given the recessive allele to their cystic fibrosis child, but neither parent shows the condition themselves ii. see Q. 2 : probability is ¼ (0.25, 1 in 4) Rr, Rr; gametes would be R, r and R, r; A = RR normal), B = Rr (carrier), C = Rr(carrier), D = rr (with cystic fibrosis) b. enzymes replace those that cannot enter the gut from the pancreas, antibiotics control the multiplication of bacteria which might colonise the mucus in the lungs c. i. show removal of viral genes (by restriction enzyme), introduction of 'correct' gene – stitch with ligase, infection of sufferer with 'correct' gene ii. they feel that the process is not 'natural' / they worry about the risks of GMOs escaping into the environment / they worry about the genes entering other, possibly harmful, organisms
	Fill in the gaps	Mendel, garden pea. Stamens, stigma. Insects. Tall, height. Homozygous, tall, dominant. 3:1. Mendel. DNA, genes, chromosomes. Nucleus.
31: VARIATION AND SELECTION	1	a. A b. A c. A d. N e. N f. N g. N h. A
	2	a. 110 b. B, because the number of individuals providing information for the mean data is greater. This means that any anomalous result will have less of an effect. c. area 1: more of the snails are in the groups with large diameter bases. The larger the diameter of the base, the tighter the limpet can attach to the rocks to avoid being dislodged by the powerful waves. d. continuous e. height / body mass / hand span

CHAPTER	QUESTION	ANSWER(S)

	3	a. 3, 2, 1
		b. i. The snails were 'protected' inside cages, so the predators couldn't reach them.
		   ii. The paint on their shells would have been more faded.
		   iii. This means that the position of the spot on the shell would not be an input (independent) variable.
		   iv. The dark colour is a better absorber of heat energy, so the snails could be more active even though the area is cool and shady.

	4	a. a desert / dry environment: it has leaves reduced in size to spines (less water loss by evaporation), it has both shallow roots and deep roots (able to absorb any water from any depth) b. it has leaves reduced in size to spines (less water loss by evaporation) / it has both shallow roots and deep roots (able to absorb any water from any depth) / it has a stem swollen with water (water storage cells) c. 1 – D, 2 – B, 3 – F, 4 – A, 5 – E, 6 - C
	Fill in the gaps	Discontinuous, blood grouping. Continuous, body mass. Discontinuous, genes, continuous, environmental. Genotype, phenotype. Phenotype, genotype, effects of environment. Mutation , crossing over, independent assortment, fertilization. Evolution, natural selection. Genes.
32: ECOLOGY AND ECOSYSTEMS	1	a. a pyramid of numbers b. beetle, small bird c. There is a loss of energy between level 2 and level 3, so there is not enough energy to support as many organisms in level 3 (also, the level 3 organisms are usually bigger). d. The level for the oak tree is shorter, since there is only one oak tree (even though this has millions of leaves).
	2	a. tertiary consumers – tawny owl; secondary consumers – willow warbler, great tit, beetle; primary consumers – winter moth larvae, oak eggar caterpillar, field mouse; producer – oak tree b. tertiary (2), secondary (14), primary (100). Check accuracy, using the 100 primary consumers to arrange the scale. c. The owls can fly to other trees for feeding, i.e.they are part of a pyramid with many oak trees at its base.

	3	a. (in order from left) L-E-S-D
		b. Number of predators
		   Amount of food available
		   Possibility of disease

	4	a. They release digestive enzymes, especially amylase, onto the dead remains. Amylase hydrolyses starch to sugars. Lipase may also be released, and hydrolyses fats to fatty acids and glycerol. The decomposers use some of the sugars and fats in respiration, and release energy and carbon dioxide.
		b. i. 1025
		   ii. active transport of ions / cell division / synthesis of large molecules
		c. i. percentage efficiency = 1.8 / 6000 x 100 = 0.03
		   ii. food input can be controlled more easily / cattle do not use up so much energy in movement
		   iii. a possibility of disease spreading quickly / difficult to remove wastes

	5	a. i. June
		   ii. April
		   iii. the leaves on the trees absorb and reflect the light
		b. i. there is more light available to them for photosynthesis
		   ii. very little light reaches them, so it would be wasteful to keep the leaves
		c. i. May – once pollination has occurred , fertilization is possible and the petals are no longer needed
		   ii. May to October shows the petal fall (fertilization) then the maximum number of fruits which fall (corresponding to dispersal)

	Crossword	ACROSS: 3, joule; 5, ecosystem; 6, quadrat; 9, consumer; 10, heterotrophy; 12, community; 14, food chain; 17, herbivore; 18, habitat; 19, omnivore DOWN: 1, autotroph; 2, producer; 4, population; 7, transect; 8, carnivore; 11, sunlight; 13, carbon, 14, food web; 15, biomass; 16, sample

CHAPTER	QUESTION	ANSWER(S)

33: THE CYCLING OF NUTRIENTS	1	a. i. carbon compounds in living animals
		    ii. C, D or E
		    iii. B
		    iv. A
		b. i. arrow labelled P from carbon dioxide to carbon compounds in plants
		    ii. carbon dioxide + water → → (light energy, chlorophyll above arrow) → glucose + oxygen

	2	a. protease b. amino acids c. Fungi d. Hypha e. Ensure oxygen concentration on x axis, uptake on y axis / axes labelled with quantities and units / points plotted clearly / join points with smooth curve f. two methods of uptake, one method of uptake is dependent on oxygen concentration: oxygen affects aerobic respiration and so energy available for active transport g. temperature (affects activity of respiratory enzymes) h. could make sure that soil is aerated / temperature is appropriate for enzymes of respiration
	3	a. Nitrogen b. X – nitrogen fixation, Y – denitrification, Z – nitrification c. nitrate d. amino acids / protein / plant hormones e. Urea / uric acid

	4	a. i. A - evaporation
		       B - transpiration
		       C - condensation
		    ii. higher elevation so lower temperature
		b. Nutrients are soluble so dissolve in run-off water

	5	a. i. Respiration
		    ii. Combustion
		b. i. Bacteria/fungi
		    ii. Water/moderate temperature
		c. i. C
		    ii. Carbon dioxide + water → oxygen + glucose
		    iii. light/solar

	Crossword	ACROSS: 1, denitrification; 3, photosynthesis; 4, carbon dioxide; 5, nitrification; 8, fixation; 9, nodules; 12, respiration; 14, fungi; 15, bacteria; 16, decomposer; 17, excretion; 18, active; 19, amino acid DOWN: 2, eutrophication; 6, enzymes; 7, leaching; 10, protein; 11, competition; 13, nitrate
34: USEFUL MICROBES	1	a. This temperature is high enough to make sure that molecules have enough energy to react together, but not high enough to denature the enzymes in the microbes. b. to mix the contents – nutrients and microbes in particular c. i. time on x axis, mass of product on y axis / labels on axes to include quantities and units / points joined in smooth curve ii. 90% of maximum is 90/100 × 3.50 = 31.5 kg. This quantity is reached at approximately 22.5 minutes.
	2	a. show steep rise in concentration between 20 – 40 minutes, close to maximum from 80 – 120 minutes b. nutrients are used up by the mould as it grows c. they are inversely related – the concentration of nutrient falls as the concentration of amoxicillin rises d. to control the growth of colonies of bacteria inside their patients, as the bacterial growth may cause disease

CHAPTER	QUESTION	ANSWER(S)

	3	a. i. in the presence of oxygen
		      ii. as a raw material for respiration to supply the energy needed for growth
		      iii. it is easier to transport (water is heavy)
		b. penicillin
		c. i. mutation
		      ii. natural selection
		d. can antiseptic kills bacteria *on the surface of* an organism (or structure, such as a table), an antibiotic kills bacteria (or slows down their reproduction) *inside* an organism

	4	a. gene cut from human DNA using restriction enzyme / bacterium containing plasmid is opened with lysozyme / plasmid is extracted / plasmid is opened with the same restriction enzyme / factor 8 gene is stitched into opened plasmid using ligase enzyme / 'new' recombined plasmid is inserted into bacterial cell in the presence of calcium ions b. it is a factor involved in the efficient clotting of blood c. human growth hormone – can help to overcome slow growth rates / insulin – can be used in the treatment of diabetes d. they are worried that the GMO might 'escape' into the environment, with unforeseen consequences / they believe that this is an unnatural process
	Crossword	ACROSS: 1, biogas; 4, sewage; 8, chymosin; 9, GMO; 10, acid; 13, bacteria; 15; mycoprotein; 16, vector; 17, restriction; 19, antibiotic; 20, lipase; 22, acetic acid; 23, alcohol; 24, lactobacillus DOWN: 1, brewing; 2, plasmid; 3, whey; 5, protease; 6, dioxide; 7, lactic; 11, carbon; 12, fermentation; 14, amylase; 15, methane; 18, gasohol; 21, yeast
	Matching terms	Distillation, fermentation, bioreactor, lactobacillus, carbon dioxide, whey, hops, yeast, alcohol, pasteurisation, chymosin, Acetobacter, sterilization, mycoprotein, bacterium
35: HUMAN IMPACT ON THE ENVIRONMENT	1	a. G b. limpet (looks triangular) c. The limpet is definitely not the most common over the whole of the area, so the transect has given misleading information. d. There are 30 limpets in five quadrats i.e. 6 per m². Total area = 10 × 10 = 100 m², so estimated population = 6 × 100 = 600. e. 30 f. no – the estimated population would still be 600 g. Yes. Perhaps the number of quadrats in the sample was not high enough.
	2	E – A – G – B – C
	3	a. Other must be 23%. Bar graph will have type of waste on x axis, percentage of total on y axis. Bars should not be touching. Check that scale makes best use of the graph grid you use. b. This means that a material can be broken down by the activities of living organisms (usually decomposers). Paper / wood / garden waste. c. methane d. paper and board + food and garden rubbish = 56, total = 77. Percentage which is biodegradable = 56/77 × 100 = 72.7%

	4	a. Diagram should show solar radiation reaching Earth's surface, being reflected away but then back towards the Earth by a layer of greenhouse gases.
		b. methane/carbon dioxide
		c. i. would rise, as ice caps melt more quickly
		      ii. would spread further North and South as temperature will be higher further from the Equator and insects can breed in these warmer areas
		      iii. become more severe, as heat evaporates water from the seas and gradients of temperature (which cause winds) become much steeper

	5	a. i. visible light
		      ii. photosynthesis
		b. over a twenty year period there has been a rise in carbon dioxide concentration (of about 20 parts per million); each year there is a fall early in the year and a rise later in the year
		c. i. The temperature will rise, as the carbon dioxide traps the heat reflected and radiated from the Earth's surface
		      ii. burning less fossil fuel / trapping carbon dioxide in special traps beneath the Earth's surface / planting more trees to absorb carbon dioxide

CHAPTER	QUESTION	ANSWER(S)
	6	a. to supply wood for building / to provide land for crops to be grown / to provide land for domestic animals b. i. loss of trees reduces the removal of carbon dioxide during photosynthesis / there may be less fallen material for decomposers ii. soil is deprived of nutrients as the trees are removed / soil can dry out more as wind blows across it / soil can be eroded by wind and water iii. fewer breeding sites so some species will decline / fewer feeding opportunities so some species will decline / less shelter for many species
	7	a. carbon monoxide - can reduce transport of oxygen in the blood insecticides - can poison top carnivores ionising radiation - can cause mutations methane - can cause rise of global temperature sulfur dioxide - can lead to acid rain untreated sewage - can spread diseases such as cholera b. i. vehicle exhausts ii. burning of fossil fuels iii. nuclear power stations
	Fill in the gaps	Infra-red, methane, CFCs, carbon dioxide. Pests, photosynthesis. CFCs. Ultraviolet, cancer, mutation. Ozone, mesophyll. Nitrates, pesticides. Acid rain. Batteries, motor oil, CFCs.
16: FARMING AND FOOD PRODUCTION	1	INCREASE YIELD: cutting down hedges – more space for crop / easier to use machinery; increased use of nitrate fertilizers – nitrate is a limiting factor for many crop plants. DAMAGE THE ENVIRONMENT: burning stubble – smoke / carbon dioxide enters the environment; drainage of wet fields – loss of habitat for wading birds; repeated growth of same crop – removes same mineral nutrients from soil each time the crop is harvested / can increase chances of disease organisms breeding in one habitat
	2	a. approximately 9% b. female c. 5% d. 25 – 29 e. B: high proportion of children, very small proportion reach 60+ f. 2 400 000 g. Food produced more efficiently, so fewer children die and family size increases. More people survive to breeding age.
	3	a. i. potatoes – fallow – sprouts – peas ii. growing a series of different crop species so that the soil is never allowed to deteriorate (do not remove same nutrients every year / include crops which can 'fix' nitrogen) b. i. Two separate blocks of bars – peas and sprouts – on x axis, nitrate level on y axis ii. mean for sprouts = (54 + 44 + 52)/3 = 50; mean for peas = (82 + 88 + 80)/3 = 83.3. Increase = (83.3 – 50)/50 × 100 = 67% increase
	4	a. suitable temperature helps to avoid loss of energy from cattle / light can affect growth rates by making more time available for feeding. b. high protein / low fat / correct minerals and vitamins / cheap to produce / easy to distribute to animals c. to know when the optimum weight gain has taken place – do not wish to continue expensive feeding if there is no further weight gain d. contains nitrates and phosphates which could lead to eutrophication of local water sources (so oxygen concentration falls and many aerobic organisms die out) e. avoid disease (antibiotics / vaccinations) which can make animals grow more slowly
	5	a. to dissolve solutes for transport / to break down large molecules in digestion / as a raw material for photosynthesis / to lubricate dry surfaces / as support e.g. in turgor b. nitrate c. i. year on x axis – bars should not be touching / check key to distinguish oxygen from nitrate ii. percentage decrease = decrease/original value × 100, so percentage decrease = 2.0/5.9 × 100 – 33.9 d. i. eutrophication ii. the availability of nitrate and phosphate iii. protein

CHAPTER	QUESTION	ANSWER(S)
	Matching terms	A-E; B-I; C-J; D-G; E-A; F-C; G-D; H-B; I-H; J-F
37: CONSERVATION	1	a. 60 divided by 12 years = 5% per annum b. The sample is very small, and may not accurately represent the whole population. c. DDT makes eggshells thinner, so fewer eggs hatch (parents break many eggs as they incubate them) d. It is at the top of a food chain, and the DDT from organisms lower in the chain will accumulate in the body of the top predator. e. Eats insects from DDT treated trees, so probably picks up DDT from the surface of the insects as well as from the contents of the insects. It will be easy for the roller to catch insects killed / slowed down by DDT. f. DDT is now banned in the UK, but remains in use in Uganda as there are few alternative methods for insect control

	2	a. i. a smooth, moist skin
		   ii. fish / reptiles / birds / mammals
		b. They are from the same genus, *Bufo*
		c. Their favoured habitat of sand dunes is under threat – from building camp sites and from the growth of woodland trees which cool the sand surface below the temperatures the Natterjack toads prefer.
		d. restrict development of camp sites, and keep sand dunes clear of trees
		e. i. secondary consumers / carnivores
		   ii. adult insects and insect larvae
		   iii. common toad has a wider range of food types, so can adapt to a loss of one or two food types

	3	a. feathers (e.g on tail) and beak
		b. i. binomial system means two-part name (Genus first, then species) e.g. *Oxyura jamaicanensis* for the Ruddy duck
		   ii. They belong to the same genus (*Oxyura*) and they are capable of cross-breeding.
		c. i. The Ruddy duck is also found in other parts of the world
		   ii. The Ruddy duck might be a more efficient feeder than the White-headed duck and so removes food sources from the White-headed duck.
		d. There are two different food sources for the duck, and there are two predators which feed on the duck. This could not be shown in a single food chain.

	4	a. They have fur, and feed their young on milk from mammary glands b. loss of habitat e.g. for cattle ranching, so the Tamarins cannot find food and are more easily captured by predators. They breed very slowly, with small litters, and are often captured for the fur trade. c. They will be removed from the wild as they are difficult to breed, the populations will become inbred and any animals that are produced cannot learn the skills they need to survive in the wild. d. They are protected, so might be reintroduced to new habitat if it becomes available / biologists can study them and find out the best conditions for breeding them / they are attractive and might help to raise funds for conservation (Flagship species) e. CITES controls the trade of endangered species and their products between different countries e.g. ivory from elephants cannot be sold from African countries to other parts of the world f. They are attractive and might help to raise funds for conservation. Saving their habitat will help other species that live there.
	5	a. 5000 miles per year in 1968 b. The UK is able to import much of its food and so needs less land to grow crops to feed its population c. SUPPORT: i, iv, v, vi ARGUE AGAINST: ii, iii, vii, viii
	Crossword	ACROSS: 1, recreation; 4, panda; 6, food web; 8, reclamation; 9, CITES; 13, creation; 14, habitat; 16, breeding; 17, tiger; 18, resources; 19, zoo; 20, elephant DOWN: 2, captive; 3, species; 5, agriculture; 7, flagship; 10, rhinoceros; 11, management; 12, mollusc; 15, DDT

Index

Page numbers in *italics* indicate question and answer sections.

A

abiotic environments 130
absorption 42–4, *45–7*, *161–2*
 absorption in the small intestine 43
accommodation 84
acid rain 147
acrosome 104
active immunity 62
active transport 28
adrenal glands 88
aerobic respiration 70
agriculture 151, *152–3*, 154, *177–8*
AIDS 63
alcohol 80
algae 16
alimentary canal 42
alleles 112
alveoli 66
amino acids 38
 deamination 43, 74, 137
ammonium ions 137
amniocentesis 108
amnion 108
amniotic fluid 108
amphibians 16, 17
amylases 34
anabolic steroids 80
anaerobic respiration 70
animals 16, 17
 animal cell features 22
annelids 16
ante-natal care 107
anthers 97
antibiotics 80, 125, 141
antibodies 38, 57
 lymphocytes 62
anus 42
aorta 58, 74
aphids 48
arteries 57
 coronary artery 57
 renal artery 74
 umbilical artery 108
artificial insemination 106
asexual reproduction 96
assimilation 42
autotrophs 130
auxins 92
axons 78

B

bacteria 62, 125, 141, 142
baking 70, 141
basement membrane 66
bile 43
biodegradable materials 154
biogas 141
biological molecules 30, *30–3*, *159–60*
biomass 131
bioreactors 142
biotic communities 130
birds 16, 17
birth 107
bladder 74, 104
blind spot 84
blood 108

blood pressure 57, 58
blood sugar regulation 89
blood tissue 57
blood vessels 23, 44
clotting 57, 62
functions of blood cells 57
pH 66
brain function 78
 homeostasis 79
 integration centre 2
 visual centre 84
breastfeeding 39, 107
breathing *67–9*, *165*
 breathing in 66
 breathing out 66
 exercise 66
 rate and depth 66
brewing 70, 141
bronchioles 66
bronchitis 80
bubble potometers 52

C

calcium 38
cambium 53
cancer 117, 147
canine teeth 44
cap (diaphragm) 105
capillaries 57, 58
 glomerulus 74
carbohydrates 38, 136
carbon dioxide 48, 66, 74
 carbon cycle 136, 137
carnivores 130, 131
carpel 97, 100
 ovary 97
 stigma 97
 style 97
cells 22–3, *24–7*, *158–9*
 animal cell features 22
 cell division 117, 118–20
 cell membranes 22, 28, 38, 66
 cell sap 22
 cells, tissues and organ systems 23
 common features of cells 22
 egg cells 105
 movement of molecules in and out of
  cells 28, 28–9
 nerve cells 78
 plant cell features 22
 specialized cells 23
 tooth-forming cells 44
 working out the size of cells 22
central nervous system (CNS) 78, *81–3*,
 166–7
 comparing nervous system with
  endocrine system 88
 drug abuse 80
 homeostasis 79
centromeres 112
cervix 105
 mucus plug in cervix 108
chewing 44
chlorophyll 22
chloroplasts 22
cholera 42
cholesterol 38
chromosomes 112
 alleles 112

centromeres 112
 homologous pairs 112, 122
 X- and Y-chromosomes 122
cilia 66
ciliary muscles 84
circulatory system 23, 57–8, *59–61*
 double circulation 57
 structure and function in mammalian
  circulation 58
cirrhosis 43, 80
cloning 126
codominance sex linkage 122
coil (I.U.D.–Intra-Uterine Device) 105
collagen fibres 44
combustion 136
communities (biotic communities) 130
condensation 136
condoms 104
cones 84
conservation 154, *155–7*, *178–9*
consumers 131
continuity variables 12
contraceptive methods 104, 105, 106
cornea 84
coronary heart disease 57, 58, 80
cortex (kidneys) 74
cortex (plants) 53
cotyledons 100
crop rotation 151
cross-pollination 97
crosswords 5
crustaceans 16
cystic fibrosis (CF) 121
cytoplasm 22, 105

D

data 12
 data presentation 13
deamination 43, 74, 137
death 136, 137
decomposers 131, 136, 137
deficiency diseases 38
deforestation 147, 151
dendrites 78
desertification 39
detectors 79
diabetes 88
 type I diabetes 88, 89
 type II diabetes 88, 89
dialysis 75
diarrhoea 42
diet 38–9, *40–1*, *160–1*
 food supply and famine 39
 ideal human diet 38–9
diffusion 28, 108
digestion 42–4, *45–7*, 137, *161–2*
 enzymes and digestion 43
digestive system 42
disease defence 62–3, *64–5*, *164*
 hygienic food preparation 63
 living well 63
 personal hygiene 63
 sewage treatment 63
 waste disposal 63
DNA 17, 22
 DNA, genes and chromosomes 112,
  113–16, *171–2*
 double helix 112
 nucleotide bases 112